An Introduction to Number Theoretic Combinatorics

An Introduction to Number Theoretic Combinatorics

Richard M. Beekman

Copyright

Beekman, Richard M., 1958-
An Introduction to Number Theoretic Combinatorics

Copyright © 2016 Richard M. Beekman. Some rights reserved.
ISBN 978-1-329-99116-3

No part of this book may be reproduced for financial gain by any mechanical, photographic, or electronic process, or in the form of a phonograph recording, nor may it be stored in a retrieval system, transmitted, or otherwise copied for public or private financial gain, except as permitted under Section 107 or 108 of the 1976 United States Copyright Act, without permission from the author.

Educators, students, academic institutions, and libraries may reproduce or copy this entire document or portions of the document, without permission from the author provided that attribution is given to the author as the originator of its contents.

Printed in the United States of America

First printing 2017

This book is dedicated to Adrian, the young aeronautical engineer.

Contents

Preface	ix
Chapter 1. Enumeration Problems of Distribution and Occupancy	1
Chapter 2. The Number-Theoretic Method	13
Chapter 3. Infinite Series and Convergence Properties	29
Chapter 4. Distributions in Identical Boxes: Recurrence Relations	39
Chapter 5. Distributions in Identical Boxes with Distinct Occupancies	65
Chapter 6. Distributions in Identical Boxes with Even Color Occupancies	93
Chapter 7. Object Distributions Derived from Polynomial Identities	125
Chapter 8. Object Distributions with Symmetric Object Sets	139
Chapter 9. Distributions in Boxes of Several Types	149
Chapter 10. Distributions in Distinct Boxes	161
Chapter 11. Generalized Exponential Bell Partition Polynomials	191
Chapter 12. Distributions in Identical Boxes with Ordered Occupancies	205
Chapter 13. Algebraic Identities for Object Distribution Polynomials	223
Chapter 14. A Generalization of Polya's Enumeration Theorem	237
Appendix A. Object Distribution Polynomials	251
Appendix B. Research Problems	257
Appendix C. Solutions to the Exercises	259
Bibliography	271
Index	273

Preface

What is number theoretic combinatorics?

Number theoretic combinatorics is a way of thinking about combinatorial problems, especially enumeration problems of distribution and occupancy, using concepts, principles, and methods borrowed from number theory.

Number Theory and Combinatorics

Number theory is a branch of mathematics that studies the properties of natural numbers (counting numbers) and the integers. Combinatorics is a branch of mathematics that studies finite or countable discrete structures and configurations. We will be primarily interested in *enumerative combinatorics*, which is concerned with *counting* the sizes of certain finite sets. Number theory and combinatorics are sister disciplines. They both involve the study of discrete objects and structures. So, it should not be surprising that there is interplay between these two fields. Each field illuminates and enriches the other.

Combinatorial Number Theory

A well-established special branch of mathematics, called *combinatorial number theory*, uses combinatorial, or counting, arguments to prove theorems in number theory. Consider the following simple example:

Theorem. *The product of k consecutive positive integers is divisible by $k!$.*

Proof. For $k \geq 1$ and $r \geq 0$, $\binom{r+k}{k}$ is a positive integer, because $\binom{r+k}{k}$ is the number of combinations of $r+k$ things chosen k at a time. Then the product of the k consecutive positive integers $r+1, r+2, \ldots, r+k-1, r+k$ is divisible by $k!$ because $\binom{r+k}{k} = \frac{(r+k)(r+k-1)\ldots(r+1)}{k!} = $ a positive integer. □

This is an example of a theorem from combinatorial number theory. Most elementary textbooks neglect to mention that you can also go in the other direction. You can use principles of number theory to prove theorems in combinatorics. This is especially true for combinatorial problems of distribution and occupancy.

Problems of Distribution and Occupancy

Throughout most of this book we will be studying a special area of enumerative combinatorics, namely problems of *distribution and occupancy*. Basically, we want to count, or enumerate, the number of ways to distribute an arbitrary set of objects, such as colored balls, into a set of arbitrary boxes (urns, or cells). When the boxes are distinct, or labeled, conventional methods exist for counting the number of distributions. There are also special cases of object sets and box sets that are readily handled using standard methods. However, when the boxes, or cells, are indistinguishable (identical), the enumeration problems become quite difficult. As John Riordan said in his book, *Introduction to Combinatorial Analysis*, "When cells are alike, the general problems of distribution and occupancy are of great involvement and only the simplest results are noticed here." (See [1], p. 99.) This is where number theoretic combinatorics enters the picture.

Number Theoretic Combinatorics

Number theoretic combinatorics is the flip-side of combinatorial number theory. Instead of using combinatorial ideas to prove theorems in number theory, we use ideas from number theory to prove theorems in combinatorics. The central idea in number theoretic combinatorics is that placing objects into boxes is equivalent to placing prime numbers into boxes. This approach allows us to convert a combinatorics problem into a number theory problem.

The number theoretic paradigm: Placing objects into boxes is equivalent to placing prime numbers into boxes.

If we represent each type of object, such as a colored ball, by a unique prime number, then the entire set of objects can be represented by a *single number*, called the *specification number n*, simply by multiplying together all the primes. Since the Fundamental Theorem of Arithmetic guarantees uniqueness of prime factorization, we can, in theory, recover the set of objects (primes) by prime factorizing the specification number for the set of objects.

To make this idea work, we need a *generating function*. Generating functions are the central tools of enumerative combinatorics. They give us a mathematical object that can be formally manipulated, using algebra and formal calculus, to discover combinatorial relationships and identities. Even when we are unable to derive an explicit formula to count a set of object distributions, we can obtain a recurrence relation that allows us to systematically calculate *object distribution polynomials*. The object distribution polynomials tell us the number of ways that we can distribute objects, of some specification, into boxes of some specification. When we combine mathematical theory with computer programs, we can calculate reference tables of these object distribution polynomials. Appendix A contains tables for many of these polynomials.

Several possibilities exist for a number theoretic generating function. One possibility is the *Dirichlet series generating function*. In this text, however, we will use a new kind of generating function, invented by the author in 1995, called the *logarithmic generating function*. (The logarithmic generating function is equivalent to the Dirichlet series generating function under the transformation $x = e^{-s}$, but the logarithmic generating function is more naturally suited to our objectives.) The reason why we use the logarithmic generating function, instead of some other generating function, is that the *logarithm* function is a *homomorphism*. It readily converts addition problems into multiplication problems, and vice versa, since $\log(ab) = \log(a) + \log(b)$. This allows us to create natural parallels between integer partitions and non-unit integer factorizations.

Classically, we could solve general problems of distribution and occupancy using multivariable generating functions. For example, a set of three kinds of objects, such as *red*, *white*, and *blue* balls, could be represented by a generating function involving three variables, say x, y, and z. The main advantage of the number theoretic paradigm is that we can solve some really hard enumeration problems without having to use multivariable generating functions. The logarithmic generating function has only three variables, x, t, and n, regardless of the number and types of objects in the object set. This makes algebraic work more tractable, and we can easily prove some theorems that would be much harder to prove using a multivariable approach.

You may be tempted to ask which approach is better. Is it better to solve object distribution problems using multivariable generating functions (or other traditional generating functions, like ordinary power series or exponential generating functions)? Or is it better to solve the problems using a number theoretic generating function? This question is like asking whether it is better to climb a mountain on the north face or the south face. Each ascent has its unique benefits and shortcomings. Different challenges and different scenery are encountered for each ascent route. In mathematics, it is useful to have alternative viewpoints on things. New viewpoints give us new insights and better understanding of our fundamental problems.

Although mathematical jargon and theory can become quite technical and abstract, we have a simple objective:

> *Given an arbitrary set of objects, with type repetitions allowed, in how many different ways can we distribute the objects into an arbitrary set of boxes, with type repetitions allowed, subject to some occupancy constraints?*

In addition to solving this general problem, we can use number theoretic methods to generalize *Polya's Counting Theorem*. This will allow us to solve object distribution problems on finite structures whose symmetry can be represented by an automorphism group. One example is the *Ferris Wheel Problem*, where we want to count the number of ways to distribute an arbitrary set of objects, with type repetitions allowed, into the buckets of a Ferris wheel, and taking dihedral symmetries into account.

Mathematical ideas are fundamentally simple. Unfortunately, you can take any mathematical idea and make it more complicated than it needs to be. Some mathematicians take simple ideas and make them so abstract, and so bloated with technical jargon and symbolism, that they become nearly incomprehensible. Throughout this text, I will try to balance human insight and intuition with mathematical rigor, sometimes sacrificing rigor for the benefit of better understanding. This is a delicate balancing act, and one that I will probably lose.

This book is best suited for upper division mathematics majors or graduate students, since I will assume that the reader is already familiar with basic concepts of number theory (e.g., the Fundamental Theorem of Arithmetic, Bezout's theorem, integer partitions, and multiplicative number theoretic functions), combinatorics (e.g., the principle of inclusion-exclusion, recurrence relations, and generating functions), linear algebra, and abstract algebra (e.g., equivalence relations and group theory). Although the basic ideas are simple, a fairly high level of mathematical maturity may be required to frequently "shift gears" between a combinatorial way of thinking and the number theoretic paradigm. I will provide numerous concrete examples throughout the text, since examples often make theory clear that would otherwise be difficult to understand.

Most of the theory presented in this introductory text was developed by the author in the period from 1995 to 2016. Some of the results presented here are new theorems in mathematics. (See, for example, the Partition-Convolution Identity, Theorem 10.9, the Number Theoretic Generalization of Polya's Theorem, Theorem 14.4, and the algebraic identities in Chapter 13.) There are also many new proofs—using the number theoretic paradigm—of well-known mathematical theorems, especially theorems about integer partitions and Stirling numbers of the second kind.

Toward the end of the book, in Appendix B, a number of unsolved research problems are presented. I encourage the reader to freely borrow any of these ideas or problems for his or her own research projects.

Richard M. Beekman
St. Louis, Missouri
2016

Chapter 1. Enumeration Problems of Distribution and Occupancy

Before we introduce the number theoretic method in Chapter 2, let us survey and review some fundamental combinatorial problems of distribution and occupancy. This will provide a foundation for further developments later in the text. The problems presented in this chapter are the simplest of the combinatorial distribution and occupancy problems, and they are a convenient place to begin our explorations. There are virtually unlimited variations on these themes, with the only limitation being the human imagination. Some interesting variations can be found in *Enumerative Combinatorics* by Charalambides (see [2], Chapter 9).

Many combinatorial problems can be modeled as the distribution of objects, such as colored balls or marbles, into boxes (urns, cells). This is a convenient mathematical model, originally proposed by MacMahon [3], that is conceptually appealing. In applications, judgment is required to decide what an "object" is and what a "box" is. In chemistry, for example, the objects might be electrons, and the boxes might be allowable states that are specified by a set of quantum numbers.

A natural and fundamental mathematical problem is to count, or *enumerate*, the number of ways to distribute m objects into r boxes, subject to some set of constraints on the objects, boxes, or occupancy. The counting formulas, if they exist, vary depending on the constraints of distribution and occupancy. In this chapter we will temporarily use the notation $N(m, r)$ to denote the number of distributions of m objects into r boxes, but we will need to specify the constraints for each problem. Some distribution problems are difficult, and explicit formulas do not always exist. In these cases, we may need to rely on recurrence relations, generating functions, asymptotics, or other diabolical methods to count the number of distributions.

The most difficult distribution and occupancy problems are generally those where we distribute objects of mixed specification into boxes that are either identical or mixed. We will examine these problems later in the text when we present and develop the number theoretic method.

For each of the following problems, we will state the problem, its solution (as a theorem), a proof, and provide a visual "iconotype" for the problem. The iconotype provides a quick visual reference.

Problem 1.1. *m* identical objects, *r* identical boxes, no box empty, and single occupancy:

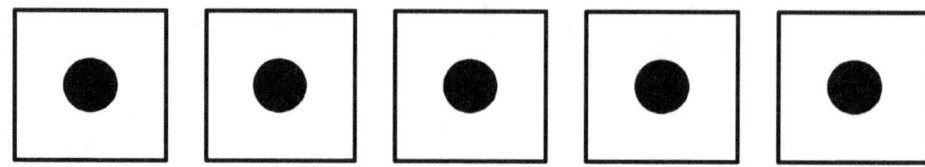

Theorem 1.1. *The number of ways to distribute m identical objects into r identical boxes, with no box empty, and single occupancy is given by* $N(m, r) = \begin{cases} 1, & m = r; \\ 0, & \text{otherwise} \end{cases}$.

Proof. In this problem, the occupancy constraints are so strict that there is either one possible distribution or no possible distributions. If the number of objects equals the number of boxes, then each box contains one object, and we are done. If the number of objects is unequal to the number of boxes, then it is impossible to place the objects into the boxes with single occupancy and no box empty. □

Problem 1.2. *m* identical objects, *r* identical boxes, no box empty, and multiple occupancy allowed:

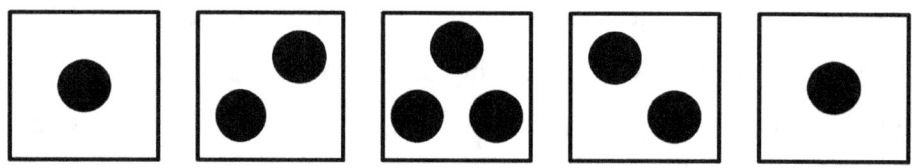

Theorem 1.2. *The number of ways to distribute m identical objects into r identical boxes, with no box empty, and multiple occupancy allowed is given by* $N(m, r) = p(m, r) = \begin{bmatrix} m \\ r \end{bmatrix}$, *where p(m, r) denotes the number of integer partitions of the positive integer m into precisely r parts.*

Proof. In this case, there is at least one object in each box. Since the boxes are identical, the order of the boxes does not matter. The number of possible distributions is the same as the number of partitions of m into precisely r parts, or $p(m, r)$. The figure above is equivalent to the integer partition $9 = 1+2+3+2+1 = 3+2+2+1+1$. A recurrence relation for the number of partitions of m into r parts is given by $p(m, r) = \sum_{k=1}^{r} p(m-r, k)$ with $m \geq r$. Many other recurrence relations are known. □

Problem 1.3. m identical objects, r identical boxes, empty boxes allowed, and single occupancy:

Theorem 1.3. *The number of ways to distribute m identical objects into r identical boxes, with empty boxes allowed, and single occupancy is given by* $N(m, r) = \begin{cases} 1, & m \leq r; \\ 0, & m > r \end{cases}.$

Proof. If the number of boxes is at least equal to the number of objects, then we can place the objects into the boxes, with single occupancy, in one way. If there are more objects than boxes, then it is impossible to place the objects into the boxes with single occupancy. □

Problem 1.4. m identical objects, r identical boxes, empty boxes allowed, and multiple occupancy allowed:

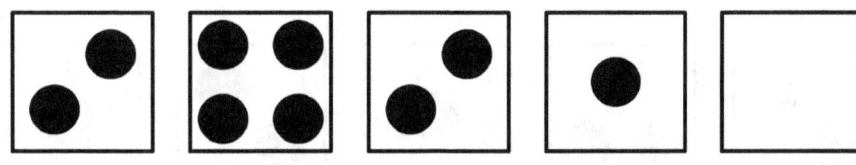

Theorem 1.4. *The number of ways to distribute m identical objects into r identical boxes, with empty boxes allowed, and multiple occupancy allowed is given by* $N(m, r) = \sum_{k=1}^{r} p(m, k) = \sum_{k=1}^{r} \begin{bmatrix} m \\ k \end{bmatrix}$, *where* $p(m, k) = \begin{bmatrix} m \\ k \end{bmatrix}$ *is the number of positive integer partitions of m into precisely k parts.*

Proof. The number of possible object distributions is equal to the number of partitions of m into at most r parts. So, we sum all the partitions of m into precisely k parts from $k = 1$ to $k = r$. □

Problem 1.5. m identical objects, r distinct boxes, no box empty, and single occupancy:

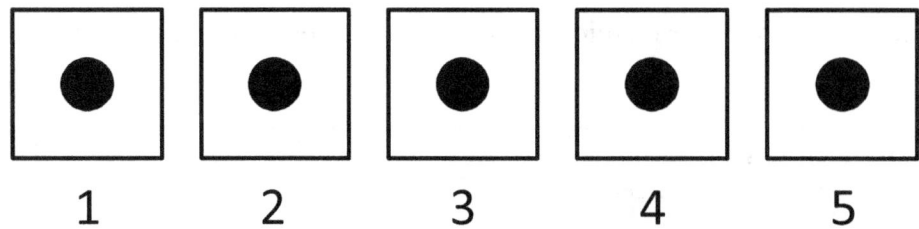

Theorem 1.5. *The number of ways to distribute m identical objects into r distinct boxes, with no box empty, and single occupancy is given by* $N(m, r) = \begin{cases} 1, & m = r; \\ 0, & \text{otherwise} \end{cases}$.

Proof. In this case, the boxes are distinct (labeled), but the objects are identical. If the number of objects equals the number of boxes, then there is precisely one way to distribute the objects into the boxes with no box empty and single occupancy. If the number of objects is unequal to the number of boxes, then no such distributions are possible. □

Problem 1.6. m identical objects, r distinct boxes, no box empty, and multiple occupancy allowed:

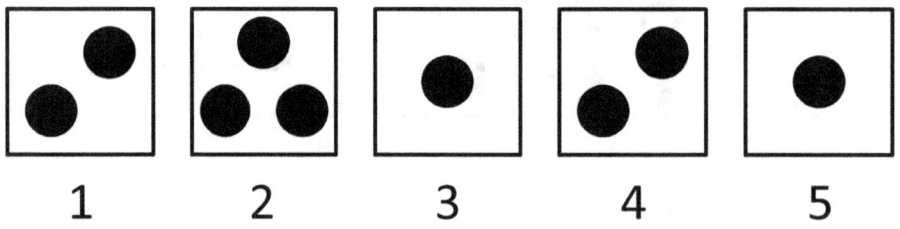

Theorem 1.6. *The number of ways to distribute m identical objects into r distinct boxes, with no box empty, and multiple occupancy allowed is given by* $N(m, r) = \binom{m-1}{r-1} = \dfrac{(m-1)!}{(r-1)!(m-r)!}$.

Proof. In this case, the boxes are distinct (labeled) and the objects are identical. Since each box must contain at least one object, we can first place one object into each box. That then leaves us with $m-r$ identical objects to effectively distribute into r distinct boxes without restriction. This problem is equivalent to Problem 1.8, below, with m replaced by $m-r$. Thus, the number of possible object distributions is $\binom{(m-r)+r-1}{m-r} = \binom{m-1}{m-r} = \binom{m-1}{r-1}$. □

Problem 1.7. m identical objects, r distinct boxes, empty boxes allowed, and single occupancy:

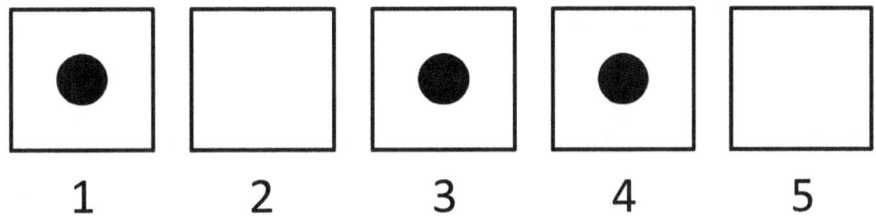

Theorem 1.7. *The number of ways to distribute m identical objects into r distinct boxes, with empty boxes allowed, and single occupancy is given by* $N(m, r) = \binom{r}{m} = \dfrac{r!}{m!(r-m)!}$.

Proof. In this case, the boxes are distinct (labeled) and the objects are identical. Since only single occupancy is allowed, placing an object into one of the labeled boxes amounts to selecting that particular box. The number of ways to choose, or select, m of the r labeled boxes is the number of combinations of r things chosen m at a time, or $\binom{r}{m}$. If $m > r$, then $\binom{r}{m} = 0$. □

Problem 1.8. m identical objects, r distinct boxes, empty boxes allowed, and multiple occupancy allowed:

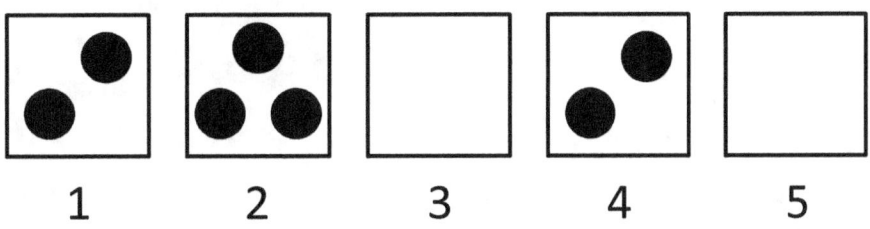

Theorem 1.8. *The number of ways to distribute m identical objects into r distinct boxes, with empty boxes allowed, and multiple occupancy allowed is given by* $N(m, r) = \binom{m+r-1}{m}$.

Proof. In this case, the boxes are distinct (labeled) and the objects are identical. Placing an object into a labeled box amounts to selecting that box from among the *r* labeled boxes. Since multiple occupancy is allowed, we are allowed to choose, or select, each box more than once. In other words, the number of possible object distributions is the number of combinations, with repetition allowed, of *r* things chosen *m* at a time. This number is $\binom{r+m-1}{m}$. □

Problem 1.9. *m* distinct objects, *r* identical boxes, no box empty, and single occupancy:

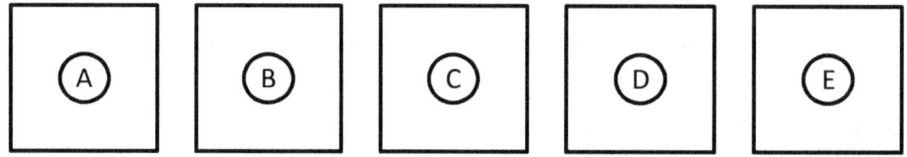

Theorem 1.9. *The number of ways to distribute m distinct objects into r identical boxes, with no box empty, and single occupancy is given by* $N(m, r) = \begin{cases} 1, & m = r; \\ 0, & \text{otherwise} \end{cases}$.

Proof. If the number of objects equals the number of boxes, then there is precisely one way to distribute the objects into the boxes with no box empty and single occupancy. If the number of objects is unequal to the number of boxes, then it is impossible to distribute the objects into the boxes subject to the given occupancy constraints. □

Problem 1.10. *m* distinct objects, *r* identical boxes, no box empty, with multiple occupancy allowed:

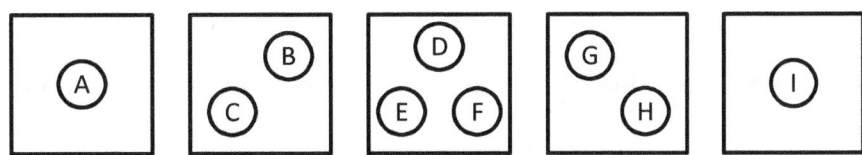

Theorem 1.10. *The number of ways to distribute m distinct objects into r identical boxes, with no box empty, and multiple occupancy allowed is given by* $N(m, r) = S(m, r) = \left\{ {m \atop r} \right\}$.

Proof. In this case, the number of possible object distributions is equivalent to the number of ways to partition an *m*-set into *r* non-empty disjoint subsets. These numbers are the *set partition numbers*, also known as the *Stirling numbers of the second kind*, denoted by $S(m, r)$ or $\left\{ {m \atop r} \right\}$. The Stirling numbers of the second kind are defined to be zero if $r > m$. A recurrence relation for Stirling numbers of the second kind is $S(n, k) = S(n-1, k-1) + k \cdot S(n-1, k)$ for $2 \leq k \leq n-1$. Initial conditions for the recurrence are $S(n, 1) = 1$ and $S(n, n) = 1$.

Problem 1.11. *m* distinct objects, *r* identical boxes, empty boxes allowed, and single occupancy:

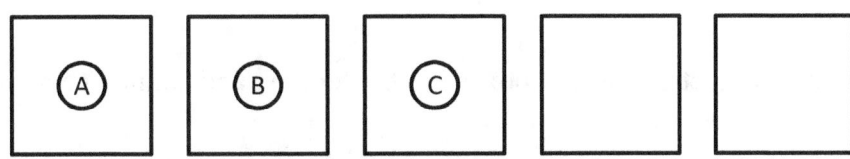

Theorem 1.11. *The number of ways to distribute m distinct objects into r identical boxes, with empty boxes allowed, and single occupancy is given by* $N(m, r) = \begin{cases} 1, & m \leq r; \\ 0, & m > r \end{cases}$.

Proof. If the number of objects is less than or equal to the number of boxes, then, with single occupancy, there is only one way to distribute the objects into identical boxes. If the number of objects is greater than the number of boxes, then it is impossible to distribute the objects into the boxes without single occupancy. □

Problem 1.12. *m* distinct objects, *r* identical boxes, empty boxes allowed, and multiple occupancy allowed:

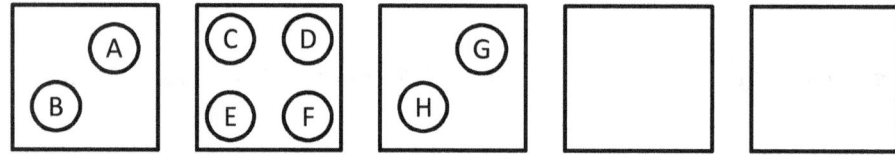

Theorem 1.12. *The number of ways to distribute m distinct objects into r identical boxes, with empty boxes allowed, and multiple occupancy allowed is given by* $N(m, r) = \sum_{k=1}^{r} S(m, k) = \sum_{k=1}^{r} \left\{ {m \atop k} \right\}$.

Proof. This case is similar to Problem 1.10, above, except that empty boxes are allowed. We can solve this problem using the method of Problem 1.10 where we first consider the number of distributions into one box, with no box empty, then into two boxes, with no box empty, then into three boxes, with no box empty, and so on. This gives us the sum of the Stirling numbers of the second kind, or $\sum_{k=1}^{r} S(m, k) = \sum_{k=1}^{r} \left\{ {m \atop k} \right\}$, where *k* is the number of boxes. □

Problem 1.13. *m* distinct objects, *r* distinct boxes, no box empty, and single occupancy:

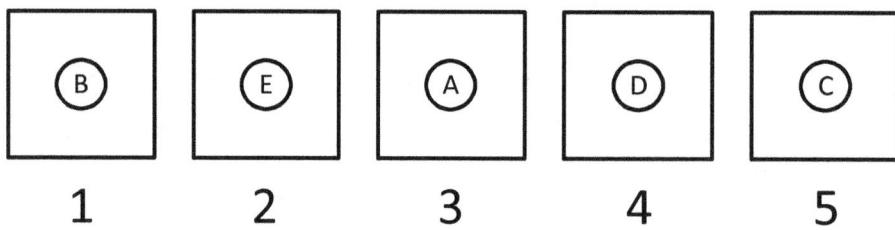

Theorem 1.13. *The number of ways to distribute m distinct objects into r distinct boxes, with no box empty, and single occupancy is given by* $N(m, r) = \begin{cases} m!, & m = r; \\ 0, & \text{otherwise} \end{cases}$.

Proof. If the number of objects equals the number of boxes, then a distribution of the objects into the boxes is effectively a permutation of the m objects. And there are $m!$ such permutations. If the number of objects does not equal the number of boxes, then no distributions are possible subject to the constraint of no box empty and single occupancy. □

Problem 1.14. m distinct objects, r distinct boxes, no box empty, and multiple occupancy allowed:

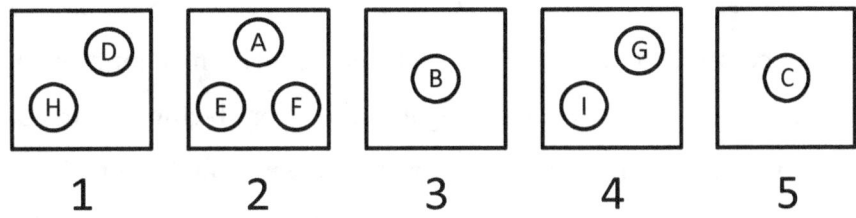

Theorem 1.14. *The number of ways to distribute m distinct objects into r distinct boxes, with no box empty, and multiple occupancy allowed is given by* $N(m, r) = S(m, r) \cdot r! = \begin{Bmatrix} m \\ r \end{Bmatrix} r!$.

Proof. First, pretend like the boxes are unlabeled. Then as in Problem 1.10, there are $S(m, r)$ possible distributions. Since the objects are distinct, no two boxes in such a distribution have the same occupancy. So, we can then label and permute the boxes themselves in $r!$ ways. Thus, the number of possible object distributions is $S(m, r) \cdot r! = \begin{Bmatrix} m \\ r \end{Bmatrix} r!$. □

Problem 1.15. *m* distinct objects, *r* distinct boxes, empty boxes allowed, and single occupancy:

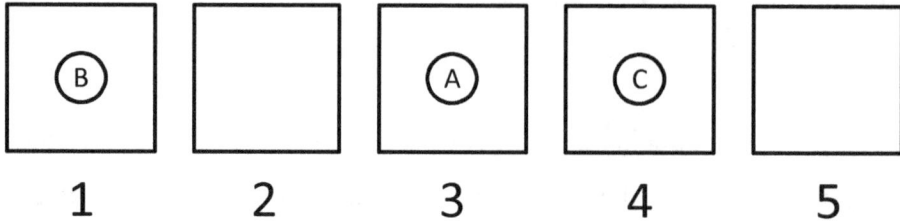

Theorem 1.15. *The number of ways to distribute m distinct objects into r distinct boxes, with empty boxes allowed, and single occupancy is given by* $N(m, r) = (r)_m = r(r-1)...(r-m+1)$ *for* $r \geq m$.

Proof. Assume the objects are labeled, say *A*, *B*, *C*, and so on. The boxes are also labeled, or numbered. Object *A* can be placed into any of the *r* labeled boxes. Since multiple occupancy is not allowed, once we select a box, we cannot select it again. So, after placing the first object *A*, we have $r-1$ choices for object *B*. We then have $r-2$ choices for object *C*, and so on. The total number of possible object distributions is $(r)_m = r(r-1)...(r-m+1)$. Clearly, the number of boxes must be greater than or equal to the number of objects. □

Problem 1.16. *m* distinct objects, *r* distinct boxes, empty boxes allowed, and multiple occupancy allowed:

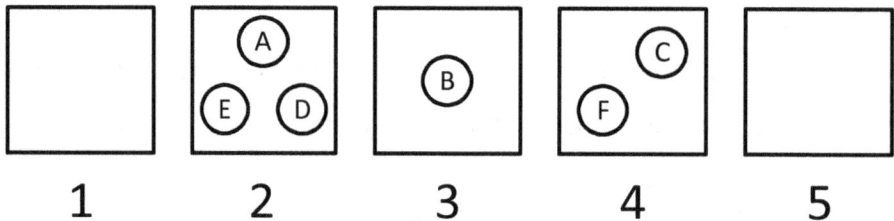

Theorem 1.16. *The number of ways to distribute m distinct objects into r distinct boxes, with empty boxes allowed, and multiple occupancy allowed is given by* $N(m, r) = r^m$.

Proof. This problem is similar to Problem 1.15, except that multiple occupancy is allowed. We have r box choices to place object A, r choices for object B, r choices for object C, and so on. Thus, there are r^m possible choices for placing m labeled objects into r labeled boxes. □

Table 1 summarizes the results for the 16 problems. Some of the problems were trivial, but they were included so that we could fill in a complete table.

Table 1. Enumeration formulas for counting the number of distributions of objects into boxes.

		\multicolumn{8}{c}{r Boxes}							
		\multicolumn{4}{c}{Identical}	\multicolumn{4}{c}{Distinct}
		\multicolumn{2}{c}{No Box Empty}	\multicolumn{2}{c}{Empty Allowed}	\multicolumn{2}{c}{No Box Empty}	\multicolumn{2}{c}{Empty Allowed}
		Single Occupancy	Multiple Occupancy	Single Occupancy	Multiple Occupancy	Single Occupancy	Multiple Occupancy	Single Occupancy	Multiple Occupancy
m Objects	Identical	Problem 1.1	Problem 1.2 surjective	Problem 1.3 injective	Problem 1.4 unrestricted	Problem 1.5	Problem 1.6 surjective	Problem 1.7 injective	Problem 1.8 unrestricted
		$1, m=r;$ $0, \text{otherwise}$	$p(m,r) = \begin{bmatrix} m \\ r \end{bmatrix}$	$1, m \leq r;$ $0, m > r$	$\sum_{k=1}^{r} \begin{bmatrix} m \\ k \end{bmatrix}$	$1, m=r;$ $0, \text{otherwise}$	$\binom{m-1}{r-1}$	$\binom{r}{m}$	$\binom{m+r-1}{m}$
	Distinct	Problem 1.9	Problem 1.10 surjective	Problem 1.11 injective	Problem 1.12 unrestricted	Problem 1.13	Problem 1.14 surjective	Problem 1.15 injective	Problem 1.16 unrestricted
		$1, m=r;$ $0, \text{otherwise}$	$S(m,r)$	$1, m \leq r;$ $0, m > r$	$\sum_{k=1}^{r} S(m,k)$	$m!, m=r;$ $0, \text{otherwise}$	$S(m,r) \cdot r!$	$(r)_m$	r^m

The Holy Grail: A Simple Formula for All Problems

As we study combinatorial problems of distribution and occupancy, you will undoubtedly notice that a variety of methods are used to obtain a bewildering collection of theorems, formulas, and recurrence relations. You may wonder why we don't find the *Holy Grail*—a single, simple formula that counts all possible object distributions into all possible box specifications. The answer to this question is that a simple general formula probably does not exist. Why? If such a formula did exist, then, as a special case, we would have a simple formula for counting integer partitions (a special kind of object distribution problem). However, the known formula for enumerating integer partitions is not simple (Andrews [4], Theorem 5.1):

$$p(n) = \frac{1}{\pi\sqrt{2}} \sum_{k=1}^{\infty} \sqrt{k} A_k(n) \frac{d}{dn}\left(\frac{1}{\sqrt{n-\frac{1}{24}}} \sinh\left(\frac{\pi}{k}\sqrt{\frac{2}{3}\left(n-\frac{1}{24}\right)} \right) \right).$$

Exercises

1. Prove the following recurrence relation for Stirling numbers of the second kind:
   $S(n, k) = S(n-1, k-1) + k \cdot S(n-1, k)$ for $2 \le k \le n-1$, $S(n, 1) = 1$, and $S(n, n) = 1$.

2. Explain why the problem of distributing m identical objects into r identical boxes, with no box empty, is equal to the number of partitions of m into r parts.

3. In the proof of Theorem 1.6, we stated that $\binom{m-1}{m-r} = \binom{m-1}{r-1}$. Why is this identity true?

4. Calculate $S(4, 2)$, then list the set partitions for the set $\{1, 2, 3, 4\}$.

Chapter 2. The Number-Theoretic Method

Chapter 1 presented an overview and summary of several simple object distribution problems and their solutions. Now we will begin our study of object distribution problems that are more difficult, starting with the distribution of an arbitrary set of objects into identical, or indistinguishable, boxes with no box empty. We will approach this problem from a number theoretic perspective by establishing an equivalence between these object distributions and unordered factorizations of a positive integer into non-unit factors. This is the foundation of the number theoretic method. Since the foundational ideas are important for the remainder of the text, we will begin slowly and gradually to build up our mathematical intuition. A key result is the logarithmic generating function, which will be our basic tool for the remaining chapters of the book. In the following chapters, we will use clever tricks to beat the secrets out of this generating function and solve many interesting and difficult problems.

The problem that we want to solve in this chapter can be stated as follows:

> *Given n_1 balls of one color, n_2 balls of a second color, n_3 balls of a third color, ... , and n_k balls of a k-th color, in how many different ways can we distribute the colored balls into r identical boxes with no box empty?*

It is generally a good idea in mathematics to start with simple examples first. Suppose that we have two red balls and one blue ball. In how many ways can we distribute these objects into identical boxes with no box empty? Let us build the solutions graphically. For one box, we have one possible placement of the balls:

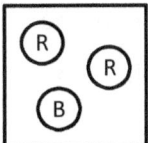

We have two possible distributions of the colored balls into two identical boxes (a dotted line surrounds each distribution):

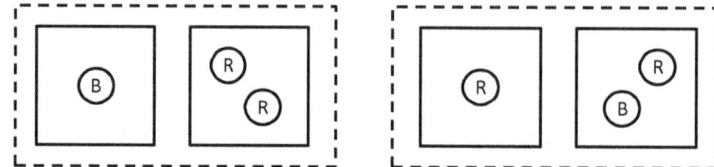

Note that since the boxes are identical, the order of the boxes does not matter. Also, the order of the objects within each box does not matter. For three identical boxes, there is only one way to place the colored balls into three identical boxes with no box empty:

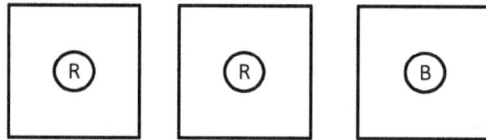

A compact notation for representing these possible object distributions is to build a polynomial—an *object distribution polynomial*—that counts the number of ways to distribute the objects into identical boxes with no box empty. For the example just shown, the polynomial looks like this:

$$g(red, red, blue, t) = t + 2t^2 + t^3.$$

The coefficient of t^k in the polynomial is the number of ways to distribute the multiset of objects $S = \{red, red, blue\}$ into precisely k boxes with no box empty. If we want to know the total number of possible object distributions, then we simply let $t = 1$ in the object distribution polynomial. Here, $g(red, red, blue, 1) = 1 + 2 + 1 = 4$. Later, we will define a more compact notation for object distribution polynomials that will facilitate algebraic computation.

Let us now give our pictures a *number theoretic* interpretation. Instead of thinking of colored balls, let us think instead in terms of *prime numbers*. By the Fundamental Theorem of Arithmetic, the prime factorization of every positive integer greater than one is unique, apart from the order of the prime factors. We can use this theorem in our present situation. Suppose that we replace the red ball by a prime number, say 2, and replace the blue ball by a different prime number, say 3. Then the set of colored balls can be represented as a multiset of prime numbers: $S = \{red, red, blue\} = \{2, 2, 3\}$. We can now reformulate our previous pictures of the object distributions. For one box, we have:

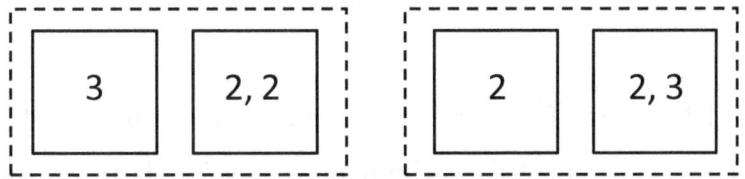

For two boxes, we have two possibilities:

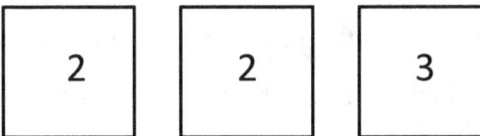

Finally, for three identical boxes, we have one possible distribution:

Now, within each box we can multiply the primes to form a positive integer. Let us do that and stack all of the four possible distributions in rows. Each row represents one of the possible object distributions (next page):

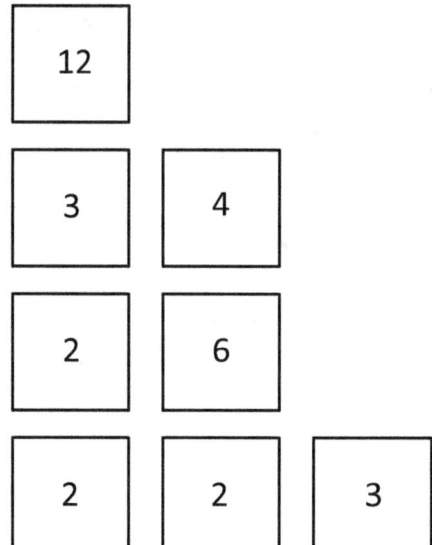

Since the Fundamental Theorem of Arithmetic guarantees the uniqueness of prime factorization, we can also reverse the process. Given the boxes with numbers shown above, we can prime factorize the numbers in each box to recapture the original prime numbers. It is easy to see that the placement of colored balls into identical boxes is equivalent to the placement of prime numbers into identical boxes. Furthermore, we see that numbers in each set (horizontal row) of boxes are the *unordered factorizations* of the number 12, which has the prime factorization $12 = 2 \cdot 2 \cdot 3$. For example, in the second row we have the unordered factorization of 12 as $3 \cdot 4$. Thus, we have the following reformulation of the object distribution problem:

> *Given n_1 balls of one color, n_2 balls of a second color, n_3 balls of a third color, ... , and n_k balls of a k-th color, form the number $n = p_1^{n_1} p_2^{n_2} p_3^{n_3} ... p_k^{n_k}$, for distinct primes $p_1, p_2, p_3, ... , p_k$. Then the number of ways that we can distribute the colored balls into r identical boxes with no box empty is equal to the number of unordered factorizations of n into precisely r non-unit factors.*

Note that the positive integer n encodes the entire set of colored balls (objects) as a single number! This gives us our first definition.

Definition 2.1. A multiset of objects, such as colored balls of k different colors, $S = \{p_1, \ldots, p_1, p_2, \ldots, p_2, \ldots, p_k, \ldots, p_k\}$, where each colored object is represented by a distinct prime number p_j, and where each color j occurs with repetition n_j, is encoded by a *specification number* $n = p_1^{n_1} p_2^{n_2} p_3^{n_3} \ldots p_k^{n_k}$.

Example 2.2. Suppose that we have three red balls, two blue balls, one green ball, and two yellow balls. Balls having the same color are identical to each other. We can represent this collection of colored balls by the multiset $S = \{red, red, red, blue, blue, green, yellow, yellow\}$. If we assign prime numbers to the object types, or colors, such that $red = 2$, $blue = 3$, $green = 5$, and $yellow = 7$, then the multiset becomes $S = \{2, 2, 2, 3, 3, 5, 7, 7\}$. We can write this object specification set more compactly as $S' = \{2^3, 3^2, 5, 7^2\}$, even though S and S' are technically different sets. This abuse of notation will cause no problems in this text. A specification number for the set of objects is then $n = 2^3 \cdot 3^2 \cdot 5 \cdot 7^2 = 17640$. Conversely, if we are given the specification number 17640, we can, in principle, prime factorize it to recover the set of object types up to an isomorphism. We could, of course, use a different assignment of prime numbers, but the specification number for the set of objects would still have the same *form*, namely $n = p_1^3 p_2^2 p_3 p_4^2$, for distinct primes p_1, p_2, p_3, and p_4. The object distribution polynomial $g_n(t)$ will be the same polynomial for all isomorphic forms of the specification number.

Now that we have established the equivalence between object distributions into identical boxes and unordered factorizations of the specification number, n, into non-unit factors, let us derive a generating function to count these unordered factorizations (object distributions).

Euler's generating function for enumerating the partitions of a positive integer n is given by ([5], Theorem 26.3):

$$P(x) = \prod_{d=1}^{\infty} \frac{1}{(1-x^d)} = \sum_{n=0}^{\infty} p(n) x^n . \tag{2.1}$$

As an example, there are seven partitions of the number 5:

$$5 = 1+1+1+1+1$$
$$= 2+1+1+1$$
$$= 3+1+1$$
$$= 2+2+1$$
$$= 4+1$$
$$= 3+2$$
$$= 5$$

Therefore, we have $p(5) = 7$, which is indicated by the term $7x^5$ in the expansion of (2.1). The idea behind the generating function is that if you expand the infinite product on the left-hand side of equation (2.1) to get a formal power series, then the coefficient of x^n will be $p(n)$. We can use the generating function to calculate values of $p(n)$, for small values of n, or to perform theoretical analyses. We can manipulate the generating function to obtain special formulas and recurrence relations for $p(n)$.

How can we do a similar thing for unordered factorizations of n? The generating function (2.1) for partitions counts the number of ways to represent a positive integer n as an unordered *sum* of positive integers. This is an *addition* problem. We want a generating function that counts the number of ways to represent a positive integer as an unordered *product* of non-unit positive integers. That is a *multiplication* problem. Fortunately, we have a clever trick. Since $\log(a) + \log(b) = \log(ab)$, we can use logarithms to convert an "addition" problem into a "multiplication" problem. By using logarithms we can convert Euler's generating function (2.1) into a generating function for unordered factorizations in non-unit factors. To do this, we replace x^d with $x^{\log d}$. It doesn't matter which logarithm base we use, so we will use the natural (base e) logarithm. Note that $x^{\log r}$ multiplied by $x^{\log s}$ gives $x^{\log r} x^{\log s} = x^{\log r + \log s} = x^{\log(rs)}$. Hence, we can use the generating function for partitions, $P(x)$, to find the number $g(n)$ of unordered non-unit factorizations of n provided that we make two modifications to $P(x)$. First, we must replace x^d with $x^{\log d}$. Second, since we do not want unit factors (factors equal to 1), we cannot allow d to be 1. If we make these substitutions into Euler's generating function (2.1), we immediately obtain the following result:

Theorem 2.3 (Logarithmic Generating Function). *Let $g(n)$ denote the number of unordered factorizations, excluding unit factors, of the positive integer $n > 1$. A generating function for $g(n)$ is given by $G(x) = \prod_{d=2}^{\infty} \dfrac{1}{(1-x^{\log d})} = \sum_{n=1}^{\infty} g(n) x^{\log n}$. It also follows from this equation that $g(1) = 1$.*

Note that $\dfrac{1}{(1-u)} = 1 + u + u^2 + u^3 + \ldots$ is the formal power series for $\dfrac{1}{(1-u)}$.

Example 2.4. Let us use Theorem 2.3 to find $g(4)$. We have

$$G(x) = \prod_{d=2}^{\infty} \frac{1}{(1-x^{\log d})}$$
$$= (1 + x^{\log 2} + x^{2\log 2} + \ldots)(1 + x^{\log 3} + x^{2\log 3} + \ldots)(1 + x^{\log 4} + x^{2\log 4} + \ldots) \ldots$$
$$= (1 + x^{\log 2} + x^{\log 2^2} + \ldots)(1 + x^{\log 3} + x^{\log 3^2} + \ldots)(1 + x^{\log 4} + x^{\log 4^2} + \ldots) \ldots$$
$$= (1 + x^{\log 2} + x^{\log 4} + \ldots)(1 + x^{\log 3} + x^{\log 9} + \ldots)(1 + x^{\log 4} + x^{\log 16} + \ldots) \ldots$$
$$= 1 + x^{\log 2} + x^{\log 3} + 2x^{\log 4} + x^{\log 5} + 2x^{\log 6} + x^{\log 7} + 3x^{\log 8} + \ldots$$

Since the coefficient of $x^{\log 4}$ is 2, we have $g(4) = 2$. This means that there are two unordered factorizations of 4 into non-unit factors (4 and $2 \cdot 2$). In terms of object distributions, if we represent a red ball by the prime number 2, then the multiset of objects $S = \{red, red\} = \{2, 2\}$ has specification number $n = 2 \cdot 2 = 4$. Then, since $g(4) = 2$, we know that there are two ways to distribute two red balls into identical boxes with no box empty. We can put both red balls into one box. Or we can put a red ball into each of two boxes. We will obtain more efficient ways of extracting the coefficients $g(n)$ in the next chapter when we derive some recurrence relations.

We will often need to multiply two logarithmic generating functions, so it is useful to know what happens to the coefficients of $x^{\log n}$.

Theorem 2.5 (A Cauchy-Type Product Rule). *The product of two logarithmic generating functions* $\sum_{n=1}^{\infty} g(n) x^{\log n}$ *and* $\sum_{n=1}^{\infty} h(n) x^{\log n}$ *is given by*

$$\sum_{n=1}^{\infty} g(n) x^{\log n} \cdot \sum_{n=1}^{\infty} h(n) x^{\log n} = \sum_{n=1}^{\infty} \left(\sum_{d|n} g(d) h(n/d) \right) x^{\log n}$$

$$= \sum_{n=1}^{\infty} \left(g(n) * h(n) \right) x^{\log n}.$$

The quantity $g(n) * h(n) = \sum_{d|n} g(d) h(n/d)$, where the summation is taken over all positive integer divisors d of n, is called the *convolution product*. The convolution product is a commutative product. It is very important in number theory, and we will use it frequently.

Proof.

$$\sum_{n=1}^{\infty} g(n) x^{\log n} \cdot \sum_{n=1}^{\infty} h(n) x^{\log n} = \sum_{1 \leq j, k} g(j) h(k) x^{\log j + \log k}$$

$$= \sum_{1 \leq j, k} g(j) h(k) x^{\log(jk)}$$

$$= \sum_{m=1}^{\infty} \sum_{1 \leq j : j | m} g(j) h(m/j) x^{\log m}, \text{ where } m = j \cdot k$$

$$= \sum_{m=1}^{\infty} \left(\sum_{j|m} g(j) h(m/j) \right) x^{\log m}$$

$$= \sum_{n=1}^{\infty} \left(\sum_{d|n} g(d) h(n/d) \right) x^{\log n}, \text{ by change of variables}$$

$$= \sum_{n=1}^{\infty} \left(g(n) * h(n) \right) x^{\log n}. \quad \square$$

Example 2.6. Let's multiply two series and see what happens:

$$\sum_{n=1}^{\infty} a_n x^{\log n} \cdot \sum_{n=1}^{\infty} b_n x^{\log n} = (a_1 x^{\log 1} + a_2 x^{\log 2} + a_3 x^{\log 3} + a_4 x^{\log 4} + \ldots)(b_1 x^{\log 1} + b_2 x^{\log 2} + b_3 x^{\log 3} + b_4 x^{\log 4} + \ldots)$$

$$= (a_1 b_1) x^{\log 1} + (a_1 b_2 + a_2 b_1) x^{\log 2} + (a_1 b_3 + a_3 b_1) x^{\log 3} + (a_1 b_4 + a_2 b_2 + a_4 b_1) x^{\log 4} + \ldots$$

$$= \left(\sum_{d|1} a_d b_{\frac{1}{d}} \right) x^{\log 1} + \left(\sum_{d|2} a_d b_{\frac{2}{d}} \right) x^{\log 2} + \left(\sum_{d|3} a_d b_{\frac{3}{d}} \right) x^{\log 3} + \left(\sum_{d|4} a_d b_{\frac{4}{d}} \right) x^{\log 4} + \ldots$$

$$= (a_1 * b_1) x^{\log 1} + (a_2 * b_2) x^{\log 2} + (a_3 * b_3) x^{\log 3} + (a_4 * b_4) x^{\log 4} + \ldots$$

This result agrees with Theorem 2.5.

Linear Independence

The generating function in Theorem 2.3 is a formal mathematical object. We are not usually concerned with its *analytic* properties. Instead, we view the generating function as an *algebraic* object—a formal power series whose coefficients are the functions $g(n)$. Thus, Theorem 2.3 gives us an identity in a ring of formal power series, and the variable x is just a formal symbol that need not take on any particular value. (If we *do* allow x to assume real or complex values, then we must consider the analytic properties, such as convergence, of the infinite series.)

For any positive integer n, the set of formal functions $\{x^{\log 1} = 1, x^{\log 2}, x^{\log 3}, \ldots, x^{\log n}\}$ form a linearly independent set. This is what allows us to equate coefficients of $x^{\log n}$.

Theorem 2.7 (Linear Independence). *For any positive integer n, the set of functions* $\{1, x^{\log 2}, x^{\log 3}, \ldots, x^{\log n}\}$ *are linearly independent for real* $x \neq 0$.

Proof. To make the method clear, we will prove that the functions $\{1, x^{\log 2}, x^{\log 3}, x^{\log 4}\}$ are linearly independent. It will be clear, however, that the method is entirely general for all n. First, note that 1 by itself is trivially linearly independent, since $c \cdot 1 = 0$ implies that $c = 0$.
Now suppose that we have

$$c_1 \cdot 1 + c_2 x^{\log 2} + c_3 x^{\log 3} + c_4 x^{\log 4} = 0 \qquad (2.2)$$

Taking the derivative of (2.2), with respect to x, and then multiplying through by $x \neq 0$ (i.e., $x\frac{d}{dx}(\cdot)$ operation) gives

$$c_2 \cdot \log 2 \cdot x^{\log 2} + c_3 \cdot \log 3 \cdot x^{\log 3} + c_4 \cdot \log 4 \cdot x^{\log 4} = 0$$

We can now factor out $x^{\log 2}$ to obtain

$$x^{\log 2}(c_2 \cdot \log 2 + c_3 \cdot \log 3 \cdot x^{\log(3/2)} + c_4 \cdot \log 4 \cdot x^{\log(4/2)}) = 0$$

Assuming that $x \neq 0$, we can divide both sides of this equation by $x^{\log 2}$ to obtain

$$c_2 \cdot \log 2 + c_3 \cdot \log 3 \cdot x^{\log(3/2)} + c_4 \cdot \log 4 \cdot x^{\log(4/2)} = 0 \qquad (2.3)$$

Repeating the $x\dfrac{d}{dx}(\cdot)$ operation on (2.3), and multiplying through by nonzero x, gives

$$c_3 \cdot \log 3 \cdot \log(3/2) x^{\log(3/2)} + c_4 \cdot \log 4 \cdot \log(4/2) x^{\log(4/2)} = 0$$

We can now factor out $x^{\log(3/2)}$ to obtain

$$x^{\log(3/2)}(c_3 \cdot \log 3 \cdot \log(3/2) + c_4 \cdot \log 4 \cdot \log(4/2) x^{\log(4/3)}) = 0$$

Assuming that $x \neq 0$, we can divide both sides of this equation by $x^{\log(3/2)}$ to obtain

$$c_3 \cdot \log 3 \cdot \log(3/2) + c_4 \cdot \log 4 \cdot \log(4/2) x^{\log(4/3)} = 0 \qquad (2.4)$$

Applying the $x\dfrac{d}{dx}(\cdot)$ operation to (2.4), and multiplying through by nonzero x, gives

$$c_4 \cdot \log 4 \cdot \log(4/3) x^{\log(4/3)} = 0$$

Assuming that $x \neq 0$ and dividing through by $x^{\log(4/3)}$ gives

$$c_4 \cdot \log 4 \cdot \log(4/3) = 0 \qquad (2.5)$$

It is now immediately evident from (2.5) that $c_4 = 0$. But if $c_4 = 0$, then equation (2.2) reduces to

$$c_1 \cdot 1 + c_2 x^{\log 2} + c_3 x^{\log 3} = 0 \qquad (2.6)$$

We can now repeat the previous argument sequence on (2.6) to show that $c_3 = 0$, and so on, to ultimately obtain $c_1 = c_2 = c_3 = c_4 = 0$. □

Because of the linear independence of the functions $\{1, x^{\log 2}, x^{\log 3}, ..., x^{\log n}\}$, we can equate the coefficients of $x^{\log n}$ for any truncated power series in these functions. In other words, if we have $\sum_{n=1}^{m} f_n x^{\log n} = \sum_{n=1}^{m} g_n x^{\log n}$, then $\sum_{n=1}^{m} f_n x^{\log n} - \sum_{n=1}^{m} g_n x^{\log n} = \sum_{n=1}^{m} (f_n - g_n) x^{\log n} = 0$. Since the functions $\{1, x^{\log 2}, x^{\log 3}, ..., x^{\log m}\}$ are linearly independent, we must have $(f_n - g_n) = 0$, or $f_n = g_n$.

We will be working with infinite sums, where m tends to infinity. This is only a minor inconvenience, because for any *real* counting problem, we must necessarily truncate the series at a suitable large, but finite, number of terms. We can then legitimately equate coefficients on both sides of the equality to solve the counting problem.

Enumeration of Factorizations by Number of Parts

Suppose that we want to distribute colored balls, some of which may be the same color, into a specific number of identical boxes, with no box empty. Employing a generating function technique first employed by Euler, we replace each term $(1+ x^{\log d} + x^{2\log d} + x^{3\log d} +...)$ by the term $(1+ tx^{\log d} + t^2 x^{2\log d} + t^3 x^{3\log d} +...)$. The variable t serves as a "counter" variable. It counts the number of times that the factor d appears. For example, look at the term $t^2 x^{2\log d}$. Of course, $2\log d = \log d^2 = \log(d \cdot d)$, so the exponent "2" in t^2 indicates that d occurs twice, as d times d, in the exponent of x. Now suppose that we multiply an arbitrary term $t^k x^{k\log p}$ by $t^s x^{s\log q}$. We get $t^{k+s} x^{k\log p + s\log q} = t^{k+s} x^{\log(p^k q^s)}$. The exponent "$k+s$" in t^{k+s} indicates that the product under the logarithm in the exponent of x, which is $p^k q^s$ is a product having $k+s$ factors in p and q (that is, p occurs k times and q occurs s times). Thus, the variable t counts the number of unordered *factors* in n. Therefore, by inserting a counter variable t into the generating function in Theorem 2.3, we have the following important modification:

Theorem 2.8 (Logarithmic Generating Function for Object Distribution Polynomials). *The number of unordered factorizations (excluding unit factors) of the positive integer $n > 1$ is counted by the polynomials $g_n(t)$, where* $G(x, t) = \prod_{d=2}^{\infty} \frac{1}{(1-tx^{\log d})} = \sum_{n=1}^{\infty} g_n(t) x^{\log n}$. *Equivalently, if a set of objects, such as colored balls, has specification number n, then the polynomial $g_n(t)$ is the object distribution polynomial that counts the number of ways to distribute the objects into identical boxes with no box empty. The coefficient of t^k in $g_n(t)$ is the number of ways to distribute objects of specification n into precisely k identical boxes with no box empty. Note that* $\frac{1}{(1-tx^{\log d})} = 1 + tx^{\log d} + t^2 x^{2\log d} + t^3 x^{3\log d} + ...$.

(Appendix A lists many of these polynomials $g_n(t)$.)

Example 2.9. Suppose that we want to find the number of unordered factorizations of $n = 12$ in non-unit factors. Using Theorem 2.8, we only need to consider terms up to and including $d = 12$, because the number 12 has no divisors greater than itself. So, if we first multiply the product of factors $(1+tx^{\log 2}+t^2x^{2\log 2}+...)(1+tx^{\log 3}+t^2x^{2\log 3}+...)...(1+tx^{\log 12}+t^2x^{2\log 12}+...)$, we will get, after collecting like terms, a sum of the form $\sum_n g_n(t) x^{\log n}$. The coefficient of $x^{\log 12}$ is $g_{12}(t)$ and that turns out to be a polynomial in t: $g_{12}(t) = t + 2t^2 + t^3$. The coefficient of t^k gives us the number of unordered factorizations of 12 into precisely k non-unit factors. The coefficient of t is 1, so there is just one factorization of 12 into one part: 12. The coefficient of t^2 is 2, so there are two unordered factorizations of 12 into two parts: $3 \cdot 4$ and $2 \cdot 6$. Finally, the coefficient of t^3 is 1, so there is one unordered factorization of 12 into three parts: $2 \cdot 2 \cdot 3$. We can also find the total number of unordered factorizations of 12 into non-unit parts by calculating $g_{12}(1) = 1 + 2 + 1 = 4$.

Equivalent Problems

There are several problems that are logically equivalent to each other. If we can solve one of these problems, then we can solve them all.

Distribution of Objects into Identical Boxes. The first problem to consider is the problem of counting the number of ways to distribute a set of objects, such as colored balls, into identical boxes with no box empty. For example, one way to distribute the multiset of objects, such as colored balls, given by $S = \{red, red, red, red, red, white, white, blue, blue, blue\}$ into five identical boxes is shown in Figure 2.1. As discussed previously, our challenge is to count the number of ways that we can distribute a multiset S of objects into a certain number of identical boxes with no box empty.

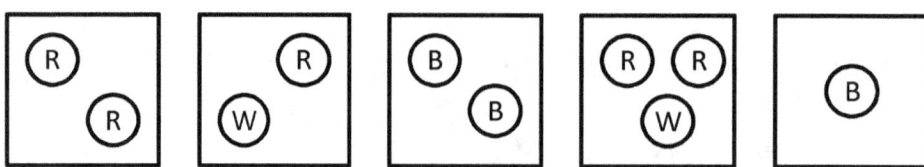

Figure 2.1. A distribution of colored balls into five identical boxes.

Unordered Factorizations of a Positive Integer. The problem of distributing a multiset of objects into identical boxes, with no box empty, is equivalent to another problem, namely the problem of determining the unordered factorizations of a positive integer into non-unit parts (i.e., no factor can be 1). To see the equivalence, simply replace the object "colors" in Figure 2.1 with prime numbers. Consider the distribution shown in Figure 2.1. We have distributed the elements (colored balls) of the multiset $S = \{red, red, red, red, red, white, white, blue, blue, blue\}$ into identical boxes. To show that this distribution can be represented as a factorization of a positive integer, we perform the following transformation. Instead of using colors to label the objects, let us use prime numbers. Let the prime number 2 represent the color *red*, let the prime 3 represent the color *white*, and let the prime 5 represent the color *blue*. We can choose any prime numbers that we want, so we might as well choose small primes. Now, when we re-label the balls for the distribution in Figure 2.1 we get the equivalent representation shown in Figure 2.2.

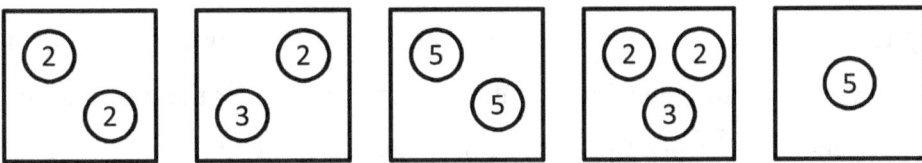

Figure 2.2. An equivalent representation of the distribution shown in Figure 2.1.

Next, multiply the numbers within each box. We then get the equivalent representation shown in Figure 2.3.

Figure 2.3. An equivalent representation of the distribution shown in Figure 2.2.

The representation shown in Figure 2.3 can be interpreted as an unordered factorization of the number 36000 in non-unit factors, namely $36000 = 4 \times 6 \times 25 \times 12 \times 5$. The order of the factors does not matter here. The next question is whether we can go the "other way" to recover the original colored balls. Given an unordered factorization into k parts with non-unit factors, like $4 \times 6 \times 25 \times 12 \times 5$, we can invoke the Fundamental Theorem of Arithmetic to uniquely prime factorize the parts: $(2^2)(2 \cdot 3)(5^2)(2^2 \cdot 3)(5)$. Thus, we recover the result of Figure 2.2: $\{2, 2\}$, $\{2, 3\}$, $\{5, 5\}$, $\{2, 2, 3\}$, and $\{5\}$. If we let the prime 2 represent the color *red*, the prime 3 represent *white*, and the prime 5 represent *blue*, then we obtain the original distribution shown in Figure 2.1:

$$\{red, red\}, \{red, white\}, \{blue, blue\}, \{red, red, white\}, \{blue\}.$$

Thus, we have a bijective mapping between the distributions of colored balls into identical boxes and the unordered factorizations of a positive integer into non-unit parts.

Vector Partitions. A *vector partition* is a partition of a vector into a sum of other vectors, without regard to the order of the summands, and where no summand is the zero vector. As an example, we have the vector partition $(7, 5) = (2, 0) + (3, 2) + (2, 3)$. This is a partition of the vector $(7, 5)$ into three non-zero parts. (The zero vector, in this case, would be $(0, 0)$.) What do vector partitions have to do with the distribution of colored balls into identical boxes? Consider the distribution in Figure 2.1 again. Since we have three different colors, *red*, *white*, and *blue*, we build a 1 x 3 vector of the form $(red, white, blue)$. Since we have five red balls, two white balls, and three blue balls, the vector is written as $(red, white, blue) = (5, 2, 3)$. Then, thinking of each box in Figure 2.1 as a "vector" of colored balls in the order $(red, white, blue)$, we have the corresponding vector partition:

$$(red, white, blue) = (5, 2, 3) = (2, 0, 0) + (1, 1, 0) + (0, 0, 2) + (2, 1, 0) + (0, 0, 1).$$

Clearly, we can reverse the process. Given a vector partition of a 1 x *m* vector into *k* summands, we can equivalently represent the vector partition as the distribution of balls of *m* different colors into *k* identical boxes.

Concluding Remarks

The general problem of distributing objects, such as colored balls, into identical boxes appears to be an intrinsically hard problem. It is almost as if there is a kind of "conservation of complexity law" at work. We cannot remove the complexity. At best, we can transfer the complexity to a more convenient place. Multivariable generating functions use one variable for each color, or kind, of object. The logarithmic generating function hides this complexity by "rolling it up" into a single specification number, n. This approach has advantages and disadvantages. As might be expected, the logarithmic generating function is not a panacea. It has its own computational challenges. However, the logarithmic generating function (Theorems 2.3 and 2.8) sometimes has advantages over other approaches:

1. The fundamental ideas are simple and easy to understand.
2. Unlike multivariable generating functions, the logarithmic generating function can easily be written in a compact form. This is a notational advantage over other methods.
3. The logarithmic generating function can easily be modified, in natural ways, to solve many "restricted" distribution problems.
4. The inherent complexity of combinatorial distribution problems is encapsulated, or hidden, within the logarithm of the specification number n.
5. The logarithmic generating function allows us to obtain new proofs of old theorems. Some of these old theorems were major achievements at the time of their original discovery. And some theorems obtained from the logarithmic generating function appear to be new to mathematics.

Exercises

1. Use Theorem 2.8 to calculate $g_{36}(t)$.
2. What does $g_{36}(t)$ mean in terms of distributing objects, such as colored balls, into boxes?
3. Explain why $g_{36}(t) = g_{100}(t)$.
4. Give a specification number n for the set of objects (colored balls) $S = \{r, r, w, b, b, g\}$. Is the specification number for this set unique?
5. Use Appendix A to find the number of ways to distribute three white balls, three red balls, and three blue balls into precisely four identical boxes with no box empty.

28

Chapter 3. Infinite Series and Convergence Properties

Throughout most of this book, we will not be concerned with the analytic properties of the logarithmic generating function. We will mostly be interested in using the generating function as a formal algebraic object to derive interesting relationships among various distribution and occupancy problems. In this chapter, we will take a brief detour to examine some of the infinite series involving the counting function $g(n)$.

The Counting Function $g(n)$

Theorem 2.3 gives us the logarithmic generating function, $G(x) = \prod_{d=2}^{\infty} \frac{1}{(1-x^{\log d})} = \sum_{n=1}^{\infty} g(n) x^{\log n}$, where $g(n)$ counts the number of unordered factorizations of a positive integer n into non-unit parts. Equivalently, $g(n)$ counts the number of ways to distribute objects of specification n (Definition 2.1) into identical boxes with no box empty.

What does the function $g(n)$ look like? First, note that $g(n)$ is an integer valued function defined on the positive integers. So, it is a discrete function. We can use the generating function $G(x)$ to calculate the first several values of $g(n)$. (Better methods will be developed later in this book.) The first several values of $g(n)$ are $g(1) = 1$, $g(2) = 1$, $g(3) = 1$, $g(4) = 2$, $g(5) = 1$, $g(6) = 2$, $g(7) = 1$, $g(8) = 3$, and $g(9) = 2$. The function appears to behave erratically. Indeed, this behavior is confirmed when we use numerical data to plot the function $g(n)$ as a function of n as shown in Figure 3.1.

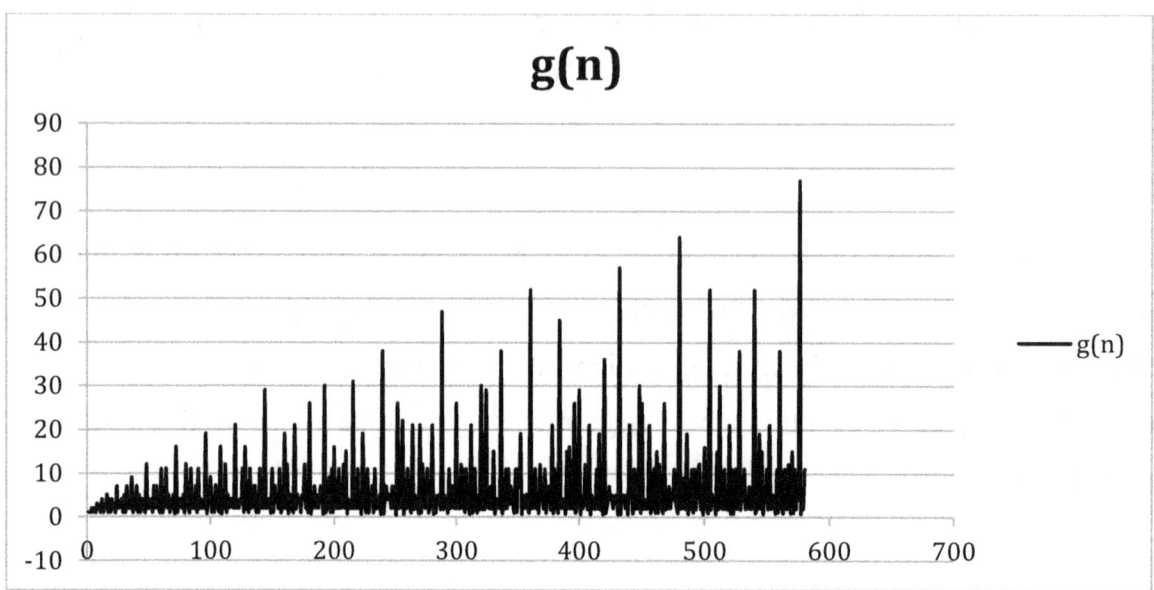

Figure 3.1. The object distribution function $g(n)$.

The parameter n is the specification number if we interpret $g(n)$ as an object distribution function. We can alternatively interpret $g(n)$ as the number of unordered factorizations of a positive integer n into non-unit factors. One of our goals is to determine an upper bound on the growth rate of $g(n)$. Theorem 3.11, by the author, shows that $g(n) \leq n^{1.7265}$.

Convergence Properties

Before we examine the growth rate of $g(n)$, let us first demonstrate some results about infinite series.

Theorem 3.1. $\sum_{n=1}^{\infty} x^{\log n}$ *diverges for real* $x \geq 1/e$, *where* $\log n$ *is the natural logarithm.*

Proof. For $x = 1/e$, we have $\sum_{n=1}^{\infty} (e^{-1})^{\log n} = \sum_{n=1}^{\infty} \frac{1}{n} = H_{\infty} \to \infty$. The series H_{∞} is the well-known *harmonic series*, which is known to diverge. For $x > 1/e$, let δ be a real number such that $\delta > 0$.

Then, by the integral test, we have $\int_1^\infty (\frac{1}{e}+\delta)^{\log u} du \to \infty$. Therefore, $\sum_{n=1}^\infty x^{\log n}$ diverges for real $x \geq 1/e$. □

Theorem 3.2. $\sum_{n=1}^\infty x^{\log n}$ *converges for* $0 < x < 1/e$.

Proof. Let $x = e^{-s}$. Then $\sum_{n=1}^\infty x^{\log n} = \sum_{n=1}^\infty (e^{-s})^{\log n} = \sum_{n=1}^\infty \frac{1}{n^s}$. But $\sum_{n=1}^\infty \frac{1}{n^s}$ converges for $\text{Re}(s) > 1$. Therefore, $\sum_{n=1}^\infty x^{\log n}$ converges for $0 < x < 1/e$ since $x = e^{-s}$. □

Theorem 3.3. *Let* $g(n)$ *be the number of ways to distribute objects of specification n into identical boxes with no box empty. (Equivalently,* $g(n)$ *is the number of unordered factorizations of n, excluding unit factors.) Then* $\sum_{n=1}^\infty g(n) x^{\log n}$ *diverges for* $x \geq 1/e$.

Proof. Since $g(n) \geq 1$ for all positive integers n, we have $x^{\log n} \leq g(n) x^{\log n}$ for all real $x > 0$. By the comparison test, we have $\sum_{n=1}^\infty x^{\log n} \leq \sum_{n=1}^\infty g(n) x^{\log n}$. But the series on the left-hand-side diverges for $x \geq 1/e$ by Theorem 3.1. Therefore, $\sum_{n=1}^\infty g(n) x^{\log n}$ also diverges for $x \geq 1/e$. □

Theorem 3.4. $\sum_{n=1}^\infty \frac{g(n)}{n}$ *diverges.*

Proof. Since $g(n) \geq 1$ for all positive integers n, by the comparison test we have $\sum_{n=1}^\infty \frac{1}{n} \leq \sum_{n=1}^\infty \frac{g(n)}{n}$. Since the left-hand-side diverges (it is the harmonic series), the right-hand-side also diverges. □

Growth Rate of $g(n)$

The next theorem, discovered by the author, is quite surprising. Theorem 3.5 gives us our first estimate for an upper bound on the growth rate of the counting function $g(n)$, namely that $g(n) = O(n^2)$. We will improve this estimate a bit in Theorem 3.11, also by the author.

Theorem 3.5. $\sum_{n=1}^{\infty} \dfrac{g(n)}{n^2} = 2.$

Proof. By Theorem 2.3, we have $G(x) = \prod_{d>1} \dfrac{1}{(1-x^{\log d})} = \sum_{n=1}^{\infty} g(n) x^{\log n}$.

Taking natural logarithms, we get $\log G(x) = -(\log(1-x^{\log 2}) + \log(1-x^{\log 3}) + \log(1-x^{\log 4}) + ...)$. Next, note that $-\log(1-y) = y + \dfrac{y^2}{2} + \dfrac{y^3}{3} + ...$, which converges for all real y such that $|y| < 1$. If we let $y = x^{\log d}$, for $d > 1$, and let $x = x_0 = e^{-2}$, then $0 < (e^{-2})^{\log d} < 1$ for all $d > 1$, and the natural logarithm series converges. Then we have

$$\log G(x_0) = \sum_{i=1}^{\infty} \dfrac{x_0^{i\log 2}}{i} + \sum_{j=1}^{\infty} \dfrac{x_0^{j\log 3}}{j} + \sum_{k=1}^{\infty} \dfrac{x_0^{k\log 4}}{k} + ...$$

$$= \sum_{i=2}^{\infty} x_0^{\log i} + \sum_{j=2}^{\infty} \dfrac{x_0^{2\log j}}{2} + \sum_{k=2}^{\infty} \dfrac{x_0^{3\log k}}{3} + ...$$

Exponentiating both sides of the equation gives

$$G(x_0) = \exp(\sum_{i=2}^{\infty} x_0^{\log i} + \sum_{j=2}^{\infty} \dfrac{x_0^{2\log j}}{2} + \sum_{k=2}^{\infty} \dfrac{x_0^{3\log k}}{3} + ...)$$

$$\sum_{n=1}^{\infty} g(n)(e^{-2})^{\log n} = \exp(\sum_{i=2}^{\infty} \sum_{k=1}^{\infty} \dfrac{1}{k \cdot i^{2k}})$$

$$\sum_{n=1}^{\infty} \dfrac{g(n)}{n^2} = \exp(\log 2)$$

$$= 2. \ \square$$

Corollary 3.6. $g(n) = O(n^2)$.

Proof. Since $\frac{g(n)}{n^2} > 0$ for all positive integers n, for any arbitrary n we have $\frac{g(n)}{n^2} < \sum_{k=1}^{\infty} \frac{g(k)}{k^2} = 2$.

Therefore, $g(n) < 2n^2$. This result also establishes an upper bound on $g(n)$. □

Theorem 3.7. Let $k \in R$, $k > 1$. Then $\sum_{n=1}^{\infty} \frac{\sqrt{g(n)}}{n \cdot k^n} \leq \sqrt{\frac{2}{k^2 - 1}}$.

Proof. By the Cauchy-Schwartz inequality and Theorem 3.5, we have

$$\sum_{n=1}^{\infty} \frac{\sqrt{g(n)}}{n} \cdot \frac{1}{k^n} \leq \left(\sum_{n=1}^{\infty} \frac{g(n)}{n^2}\right)^{1/2} \cdot \left(\sum_{n=1}^{\infty} \frac{1}{k^{2n}}\right)^{1/2}$$

$$= \sqrt{2} \cdot \sqrt{\frac{1}{k^2 - 1}} \quad \square$$

Example 3.8. Let $k = 2$ in Theorem 3.7. Then we have $\sum_{n=1}^{\infty} \frac{\sqrt{g(n)}}{n \cdot 2^n} \leq \sqrt{\frac{2}{3}} = 0.8165$. In fact, numerical data show that $\sum_{n=1}^{\infty} \frac{\sqrt{g(n)}}{n \cdot 2^n} = 0.7012$. □

The next theorem suggests an interesting conjecture. Is it true that $\sum_{n=1}^{\infty} \frac{\sqrt{3g(n)}}{n^2} = \pi$? It seems too incredible to be true, but the circumstantial evidence is suggestive. As we add more data points, the lower bound will increase, and we have $\pi = 3.1415926...$ as a ceiling. However, the rate of increase is not well understood at this time. The infinite series may converge to some real number less than π.

Theorem 3.9. $3.0469 < \sum_{n=1}^{\infty} \frac{\sqrt{3g(n)}}{n^2} \leq \pi$.

Proof. By the Cauchy-Schwartz inequality and Theorem 3.5, we have

$$\sum_{n=1}^{\infty} \frac{\sqrt{g(n)}}{n} \cdot \frac{1}{n} \le \left(\sum_{n=1}^{\infty} \frac{g(n)}{n^2}\right)^{1/2} \cdot \left(\sum_{n=1}^{\infty} \frac{1}{n^2}\right)^{1/2}$$

$$= \sqrt{2} \cdot \sqrt{\frac{\pi^2}{6}}$$

$$= \frac{\pi}{\sqrt{3}}.$$

This establishes the upper bound in the theorem. The lower bound is obtained from actual numerical data, which so far shows that $3.0469 < \sum_{n=1}^{\infty} \frac{\sqrt{3g(n)}}{n^2}$. □

If the conjecture suggested by Theorem 3.9 is true, then we should be able to show that $\sum_{n=1}^{\infty} \frac{\sqrt{3g(n)}}{n^2}$ is an irrational number. This appears to be a difficult problem, but a simple argument gives the following.

Theorem 3.10. *Either* $\sum_{n=1}^{\infty} \frac{\sqrt{g(n)}}{n^2}$ *is irrational or* $\sum_{n=1}^{\infty} \frac{\sqrt{3g(n)}}{n^2}$ *is irrational, or both.*

Proof. If $\sum_{n=1}^{\infty} \frac{\sqrt{g(n)}}{n^2}$ is irrational, then we are done. So, suppose that $\sum_{n=1}^{\infty} \frac{\sqrt{g(n)}}{n^2}$ is rational. Then there exist integers a and b, with $b \ne 0$, such that $\sum_{n=1}^{\infty} \frac{\sqrt{g(n)}}{n^2} = \frac{a}{b}$. We can now write

$$\sum_{n=1}^{\infty} \frac{\sqrt{3g(n)}}{n^2} = \sqrt{3} \sum_{n=1}^{\infty} \frac{\sqrt{g(n)}}{n^2} = \sqrt{3}\frac{a}{b}.$$

But then $\sqrt{3}\frac{a}{b}$ must be irrational, because $\sqrt{3}$ is irrational while $\frac{a}{b}$ is rational. Therefore, $\sum_{n=1}^{\infty} \frac{\sqrt{3g(n)}}{n^2}$ is irrational. □

Upper Bound on $g(n)$

First, let us discuss some preliminary observations. Let $h(n)$ denote the number of *ordered* factorizations, into non-unit factors, of the positive integer n. Clearly, we have $g(n) \leq h(n)$. This inequality follows from the observation that for each *unordered* factorization of n, into non-unit factors, counted by $g(n)$, there will be one or more *ordered* factorizations counted by $h(n)$. For example, $g(12) = 4$, but $h(12) = 8$. The four unordered factorizations of 12, into non-unit factors, counted by $g(12)$ are $12, 2\cdot 6, 3\cdot 4$, and $2\cdot 2\cdot 3$. Therefore, if we can find an upper bound on $h(n)$, say $h(n) \leq f(n)$, then we will also have an upper bound on $g(n)$: $g(n) \leq h(n) \leq f(n)$.

It is generally quite easy to calculate $h(n)$, because we are dealing with *ordered* factorizations of n. In fact, $h(n)$ can be calculated using a simple recurrence relation:

$$h(n) = \sum_{\substack{d|n \\ d>1}} h(n/d). \tag{3.1}$$

Despite the superficial similarity between $h(n)$ and $g(n)$, appearances are deceptive. While it is easy to calculate $h(n)$, it is, in general, much harder to calculate $g(n)$. Changing the word "ordered" to the word "unordered" changes the entire problem.

Theorem 3.11. $g(n) \leq n^k$, where $k = \zeta^{-1}(2) = 1.7265$ and $\zeta(s)$ is the Riemann Zeta function $\zeta(s) = \sum_{n=1}^{\infty} \frac{1}{n^s}$, for $\operatorname{Re}(s) > 1$. In other words, $g(n) \leq n^{1.7265}$.

Proof. Since $g(n) \leq h(n)$, we will prove that $h(n) \leq n^k$ for $k = \zeta^{-1}(2)$. We can prove this using mathematical induction:

<u>Induction Basis.</u> By definition, $g(1) = h(1) = 1$, even though we cannot factor the number 1 into non-unit factors. Clearly, $h(1) = 1 \leq 1^k$ for any real k. So, the result is true for $n = 1$. If you are bothered by the fact that $g(1)$ and $h(1)$ are *defined* as 1, we could start with $n = 2$. Then, $h(2) = 1 \leq 2^k$ for $k = \zeta^{-1}(2) = 1.7265$.

<u>Induction Hypothesis</u>. Suppose that $h(m) \leq m^k$ for all positive integers $m < n$, and $k = \zeta^{-1}(2)$. Then, by hypothesis, for d a non-unit divisor of n, we have $h(n/d) \leq (n/d)^k$ for $d \neq 1$. Then, by the recurrence (3.1), we have

$$\begin{aligned}
h(n) &= \sum_{\substack{d|n \\ d>1}} h(n/d) \\
&\leq \sum_{\substack{d|n \\ d>1}} (n/d)^k \\
&= n^k \sum_{\substack{d|n \\ d>1}} \frac{1}{d^k} \\
&< n^k \sum_{d>1} \frac{1}{d^k} \\
&= n^k (\zeta(k) - 1) \\
&= n^k (2 - 1) \\
&= n^k,
\end{aligned}$$

provided that we set $\zeta(k) = 2$. This implies, of course, that $k = \zeta^{-1}(2) = 1.72647$. Therefore, $h(n) \leq n^k$, and the result is true for all positive integers n. □

Corollary 3.6 and Theorem 3.11 give us upper bounds on the growth rate of the function $g(n)$. Examination of numerical data suggests that there is much room for improvement. Based on numerical data, we have the following conjecture:

Conjecture 3.12. $g(n) = O(n^{\log_e 2}) = O(n^{0.693\ldots})$.

Exercises

1. Prove that $\sum_{n=1}^{\infty} \frac{1}{k^{2n}} = \frac{1}{k^2 - 1}$ for $k > 1$. (See the proof of Theorem 3.7.)

2. Use the recurrence relation $h(n) = \sum_{\substack{d|n \\ d>1}} h(n/d)$, with $h(1) = 1$, to calculate the number of ordered factorizations of 12, or $h(12)$.

Chapter 4. Distributions in Identical Boxes: Recurrence Relations

We continue our study of object distribution problems considering the number of ways to distribute objects, of arbitrary specification, into identical boxes with no box empty. Our main goal is to develop methods for calculating the object distribution polynomials $g_n(t)$. The logarithmic generating function that was introduced in Chapter 2 (Theorem 2.8) contains all the information we need to calculate object distribution polynomials $g_n(t)$. We could, of course, simply expand the generating function, collect like terms, and read off the desired coefficient of $x^{\log n}$ to obtain the polynomial $g_n(t)$. Unfortunately, this direct approach requires too much effort. We need better ways to extract the polynomials $g_n(t)$. Ideally, we would like to obtain explicit formulas for the polynomials $g_n(t)$, but this is generally not possible. Explicit formulas only exist for special cases. Another approach, which we will develop in this chapter, is to derive recurrence relations for the polynomials $g_n(t)$. A recurrence relation is an equation that recursively defines a sequence (here, of object distribution polynomials), given one or more initial values of the sequence. Recurrence relations allow us to systematically calculate any desired object distribution polynomial. We will derive several recurrence relations. Using these recurrence relations, we can write computer programs to calculate reference tables of object distribution polynomials (see Appendix A).

Our first recurrence formula is not particularly useful by itself, but we will need it to prove a more important result (Theorem 4.2).

Lemma 4.1. *A recurrence relation for $g_n(t)$ is given by*

$$g_n(t) \cdot \log n = \sum_{d>1} \sum_{j \geq 1} g_{\frac{n}{d^j}}(t) \cdot t^j \cdot \log d$$

where d^j is a divisor of n and $g_1(t) = 1$.

Proof. By the logarithmic generating function, Theorem 2.8, $G(x,t) = \prod_{d>1} \frac{1}{(1-tx^{\log d})} = \sum_{n=1}^{\infty} g_n(t) x^{\log n}$.

Taking the natural logarithm of both sides of the equation (in reverse order) gives

$$\log\left(\sum_{n=1}^{\infty} g_n(t) x^{\log n}\right) = \log\left(\prod_{d=2}^{\infty} \frac{1}{(1-tx^{\log d})}\right)$$

$$= -\sum_{d=2}^{\infty} \log(1 - tx^{\log d})$$

Taking the derivative of both sides of the equation with respect to x gives

$$\frac{d}{dx}\left(\log\left(\sum_{n=1}^{\infty} g_n(t) x^{\log n}\right)\right) = -\frac{d}{dx}\left(\sum_{d=2}^{\infty} \log(1 - tx^{\log d})\right)$$

$$\frac{d}{dx}\left(\sum_{n=1}^{\infty} g_n(t) x^{\log n}\right) \left(\frac{1}{\sum_{n=1}^{\infty} g_n(t) x^{\log n}}\right) = -\frac{d}{dx}\left(\sum_{d=2}^{\infty} \log(1 - tx^{\log d})\right)$$

$$\frac{\sum_{n=1}^{\infty} g_n(t) \log n \cdot x^{\log n - 1}}{\sum_{n=1}^{\infty} g_n(t) x^{\log n}} = \sum_{d=2}^{\infty} t \cdot \log d \cdot x^{\log d - 1} \cdot (1 - tx^{\log d})^{-1}$$

$$\frac{\frac{1}{x} \cdot \sum_{n=1}^{\infty} g_n(t) \log n \cdot x^{\log n}}{\sum_{n=1}^{\infty} g_n(t) x^{\log n}} = \frac{1}{x} \cdot \sum_{d=2}^{\infty} t \cdot \log d \cdot x^{\log d} \cdot (1 - tx^{\log d})^{-1}$$

Multiplying through by x and rearranging, we get the following (next page):

$$\sum_{n=1}^{\infty} g_n(t) \cdot \log n \cdot x^{\log n} = (\sum_{n=1}^{\infty} g_n(t) x^{\log n})(\sum_{d=2}^{\infty} t \cdot \log d \cdot x^{\log d} \cdot (1 - t x^{\log d})^{-1})$$

$$= \sum_{n=1}^{\infty} \sum_{d=2}^{\infty} g_n(t) x^{\log n} \cdot t \cdot \log d \cdot x^{\log d} \cdot (1 - t x^{\log d})^{-1}$$

$$= \sum_{n=1}^{\infty} \sum_{d=2}^{\infty} g_n(t) x^{\log n} \cdot t \cdot \log d \cdot x^{\log d} \cdot \sum_{k=0}^{\infty} t^k x^{k \log d}$$

$$= \sum_{n=1}^{\infty} \sum_{d=2}^{\infty} \sum_{k=0}^{\infty} g_n(t) x^{\log n} \cdot t \cdot \log d \cdot x^{\log d} \cdot t^k x^{k \log d}$$

$$= \sum_{n=1}^{\infty} \sum_{d=2}^{\infty} \sum_{k=0}^{\infty} g_n(t) \cdot t^{k+1} \cdot \log d \cdot x^{\log n + \log d + \log d^k}$$

$$= \sum_{n=1}^{\infty} \sum_{d=2}^{\infty} \sum_{k=0}^{\infty} g_n(t) \cdot t^{k+1} \cdot \log d \cdot x^{\log(n d^{k+1})}$$

Now let $m = n \cdot d^{k+1}$, which gives

$$\sum_{n=1}^{\infty} g_n(t) \cdot \log n \cdot x^{\log n} = \sum_{m=2}^{\infty} \sum_{d=2}^{\infty} \sum_{k=0}^{\infty} g_{\frac{m}{d^{k+1}}}(t) \cdot t^{k+1} \cdot \log d \cdot x^{\log m}.$$

It is now to be understood that d^{k+1} is a divisor of m. Note that as k ranges from 0 to ∞, as d ranges from 2 to ∞, and as n ranges from 1 to ∞, m ranges from 2 to ∞. Since n is a dummy variable of summation on the left-hand-side of the last equation, we can replace n by m in the left-hand summation. Also, if $m = 1$, then $\log m = \log 1 = 0$. So, we have

$$\sum_{m=2}^{\infty} g_m(t) \cdot \log m \cdot x^{\log m} = \sum_{m=2}^{\infty} \sum_{d=2}^{\infty} \sum_{k=0}^{\infty} g_{\frac{m}{d^{k+1}}}(t) \cdot t^{k+1} \cdot \log d \cdot x^{\log m}$$

$$\sum_{m=2}^{\infty} (g_m(t) \log m) x^{\log m} = \sum_{m=2}^{\infty} (\sum_{d=2}^{\infty} \sum_{k=0}^{\infty} g_{\frac{m}{d^{k+1}}}(t) \cdot t^{k+1} \cdot \log d) x^{\log m}$$

Equating the coefficients of $x^{\log m}$ on both sides of the last equation gives

$g_m(t) \log m = \sum_{d=2}^{\infty} \sum_{k=0}^{\infty} g_{\frac{m}{d^{k+1}}}(t) \cdot t^{k+1} \cdot \log d$. Next, we change variables and let $j = k + 1$ to get

$g_m(t) \log m = \sum_{d>1} \sum_{j \geq 1} g_{\frac{m}{d^j}}(t) \cdot t^j \cdot \log d$. Changing variables from m to n gives the form in the theorem statement. □

Now that we have Lemma 4.1, we can prove a very useful recurrence relation for calculating a special extension of the object distribution polynomial $g_n(t)$. Suppose that we have a specification of objects, say n (Definition 2.1), whose object distribution polynomial is $g_n(t)$. We would like to calculate the object distribution polynomial $g_{nq}(t)$ for adding a single new object of a new color to the collection of objects. For example, if the original set of objects consists of *red*, *white*, and *blue* balls, with say $red = 2$, $white = 3$, $blue = 5$, then the specification number for the original set of objects has the form $n = 2^a \cdot 3^b \cdot 5^c$. The object distribution polynomial for this set of objects will be $g_n(t) = g_{2^a \cdot 3^b \cdot 5^c}(t)$. Now suppose that we add a single ball of a new color to the set of objects. Let's add a single *green* ball to the set. If we let the color *green* be represented by the prime number 7, then $green = 7$, and the specification number for the new set of objects will be $n' = n \cdot 7 = 2^a \cdot 3^b \cdot 5^c \cdot 7$, or $n' = qn$ for $q = 7$. Since the new prime q, which represents a new object of a new color, is different from the primes 2, 3, 5, q and n are necessarily coprime: $\gcd(q, n) = (q, n) = 1$. If we already know the object distribution polynomials $g_d(t)$, for each divisor d of n, how can we find the object distribution polynomial $g_{nq}(t)$? The following theorem answers this question.

Theorem 4.2. *Let n be a specification number for a set of objects. Let q be a prime number such that q is coprime to n (i.e., $(n, q) = 1$). Then the object distribution polynomial $g_{nq}(t)$ is given by*

$$g_{nq}(t) = t \cdot \sum_{d|n} g_d(t). \tag{4.1}$$

Proof. Let $m = nq$ where $(n, q) = 1$. By Lemma 4.1, we have the following recurrence relation:

$$g_m(t) \log m = \sum_{d>1} \sum_{j \geq 1} g_{\frac{m}{d^j}}(t) \cdot t^j \cdot \log d \tag{4.2}$$

Since $m = nq$, the left-hand side of (4.2) becomes

$$g_{nq}(t) \log(nq) = g_{nq}(t)(\log n + \log q) = g_{nq}(t) \log n + g_{nq}(t) \log q \tag{4.3}$$

So, from (4.3) and the fact that $(n, q) = 1$, we can find $g_{nq}(t)$ simply by looking at the coefficients of $\log q$ in the expansion of the right-hand side of (4.2). Let us denote $d_{i,n}$ to be a divisor of n. Since n and q are coprime, $d_{i,n}$ does not divide q (unless $d_{i,n} = 1$). When we expand the right-hand side of (4.2), any terms involving divisors d^j of m, where $d = d_{i,n}$, will have the form $g_{\frac{m}{d_{i,n}^j}}(t) \cdot t^j \cdot \log d_{i,n}$ and $\log d_{i,n}$ will not have a "q" in the prime factorization of $d_{i,n}$. So, these terms contribute nothing to the coefficient of $\log q$ in the expansion of the right-hand side of (4.2). The terms that do contribute to the coefficient of $\log q$ in the expansion of the right-hand side of (4.2) will have the form $g_{\frac{m}{d^j}}(t) \cdot t^j \cdot \log d$ where $d = d_{i,n}q$. Since there is only one "q" in the prime factorization of m, we must have $j = 1$ in these terms. So, the contributing terms must have the form

$$g_{\frac{m}{d_{i,n}q}}(t) \cdot t \cdot \log(d_{i,n}q) = g_{\frac{n}{d_{i,n}}}(t) \cdot t \cdot \log d_{i,n} + g_{\frac{n}{d_{i,n}}}(t) \cdot t \cdot \log q \qquad (4.4)$$

To find the total number of terms like $g_{\frac{n}{d_{i,n}}}(t) \cdot t \cdot \log q$, we note that there is precisely one such term for each divisor $d_{i,n}$ of n. Adding up all such terms, and equating coefficients of $\log q$ on both sides of equation (4.2), gives

$$\begin{aligned} g_{nq}(t) \log q &= \sum_{d|n} g_{\frac{n}{d}}(t) \cdot t \cdot \log q \\ &= t \cdot \sum_{d|n} g_d(t) \log q \end{aligned}$$

Therefore, we have $g_{nq}(t) = t \cdot \sum_{d|n} g_d(t)$. $\square$

Example 4.3. Suppose we have one *red* ball and one *blue* ball. We can represent this set of objects as $S = \{red, blue\} = \{2, 3\}$, where *red* = 2 and *blue* = 3. A specification number for this set of objects is $n = 2 \cdot 3 = 6$. You can figure out, perhaps by trial and error, that the object distribution polynomial for this set of objects is $g_6(t) = t + t^2$. Here, the coefficient of t^k in $g_6(t)$ is the number of ways to distribute one *red* ball and one *blue* ball into precisely k identical boxes with no box empty. Now suppose that we add a single *white* ball to the set of objects. What is the object distribution polynomial for this new set of objects? First, choose a new prime number, different from 2 and 3, for the color white, say *white* = 5. Our new set of objects is now $S' = \{red, blue, white\} = \{2, 3, 5\}$, and a specification number for S' is $n' = qn = 5n = 5 \cdot 2 \cdot 3 = 30$. To find the object distribution polynomial $g_{30}(t)$, we use Theorem 4.2 along with $g_1(t)$, $g_2(t)$, $g_3(t)$, and $g_6(t)$ from Appendix A. With $n = 6$ and $q = 5$, we have

$$g_{30}(t) = t \sum_{d|6} g_d(t)$$
$$= t\left(g_1(t) + g_2(t) + g_3(t) + g_6(t)\right)$$
$$= t(1 + t + t + (t + t^2))$$
$$= t(1 + 3t + t^2)$$
$$= t + 3t^2 + t^3.$$

The object distribution polynomial $g_{30}(t)$ tells us the number of ways that we can distribute three distinct objects $S' = \{red, blue, white\} = \{2, 3, 5\}$ into identical boxes with no box empty. For example, since the coefficient of t^2 in $g_{30}(t)$ is 3, we know that there are three ways to distribute three distinct objects into precisely two identical boxes with no box empty. Don't forget that since the boxes are identical, the order of the boxes does not matter.

Distinct Objects in Identical Boxes

Counting the number of distributions of distinct objects into identical boxes, with no box empty, is a relatively easy special case of the object distribution problem. This problem can be solved using standard combinatorial methods. If we have a set of distinct objects that we want to assign to identical boxes, this problem is equivalent to the set partition problem. The number of ways to partition a set of k distinct objects into precisely r identical boxes (or sets), with no box empty, is the Stirling number of the second kind, $S(k, r)$. This means that we can easily build the object distribution polynomials, $d_k(t)$, that count the number of ways to distribute k distinct objects into identical boxes with no box empty. Essentially, we have $d_k(t) = \sum_{r=1}^{k} S(k, r) t^r = \sum_{r=1}^{k} \begin{Bmatrix} k \\ r \end{Bmatrix} t^r$. Note that there are two commonly used notations for Stirling numbers of the second kind: $S(k, r)$ or $\left\{ {k \atop r} \right\}$. Both notations mean the same thing.

This problem is simple enough that we don't need to use the logarithmic generating function. However, it is instructive to show that we can use the logarithmic generating function to solve this problem if we want to. Let's first provide a couple basic definitions, then we will show how Theorem 4.2 can be used to derive a well-known theorem about Bell numbers.

Definition 4.4 (Stirling numbers of the second kind). The *Stirling numbers of the second kind*, denoted by $S(n, k)$, or $\left\{ {n \atop k} \right\}$, are the number of partitions of an *n*-set into *k* non-empty disjoint subsets, where $1 \leq k \leq n$. (See Exercise 4 in Chapter 1.)

Stirling numbers of the second kind have been extensively studied. A recurrence relation for these numbers is given by (Biggs, [5] p. 126) $S(n, k) = S(n-1, k-1) + k \cdot S(n-1, k)$, for $2 \leq k \leq n-1$, and $S(n, 1) = 1$, $S(n, n) = 1$.

Definition 4.5 (Bell numbers). The total number of partitions of an *n*-set into non-empty disjoint subsets is the *Bell number*, B_n. By definition, $B_n = \sum_{k=1}^{n} S(n, k) = \sum_{k=1}^{n} \left\{ {n \atop k} \right\}$.

Theorem 4.6. *Let $d_k(t)$ be the object distribution polynomial that counts the number of ways to distribute k distinct objects into identical boxes with no box empty. A recurrence relation for $d_k(t)$ is given by $d_{k+1}(t) = t \cdot \sum_{r=0}^{k} \binom{k}{r} d_r(t)$, with $d_0(t) = 1$ and $d_1(t) = t$.*

Proof. For a set of *k* distinct objects, the object specification number has the form $n = p_1 p_2 \ldots p_k$ for distinct primes p_j. Adding one more object, of a different kind or color, to the set gives us the specification number $np_{k+1} = p_1 p_2 \ldots p_k p_{k+1}$. By Theorem 4.2, we have $g_{nq}(t) = t \cdot \sum_{d|n} g_d(t)$, where

$g_m(t)$ is the required object distribution polynomial. So we have $g_{p_1 p_2 \cdots p_{k+1}}(t) = t \cdot \sum_{d | p_1 p_2 \cdots p_k} g_d(t)$. When we do this sum over all the divisors d of $n = p_1 p_2 \cdots p_k$, there will be $\binom{k}{r}$ possible choices for divisors of the form $d = p_1 p_2 \cdots p_r$ for any r distinct primes. And for any selection of r distinct primes, $g_d(t)$ will be the same polynomial. So we have

$$\begin{aligned} d_{k+1}(t) &= g_{p_1 p_2 \cdots p_{k+1}}(t) \\ &= t \cdot \sum_{d | p_1 p_2 \cdots p_k} g_d(t) \\ &= t \cdot \left(g_1(t) + \sum_{r=1}^{k} \binom{k}{r} g_{p_1 \cdots p_r}(t) \right) \\ &= t \cdot \left(1 + \sum_{r=1}^{k} \binom{k}{r} d_r(t) \right) \\ &= t \cdot \sum_{r=0}^{k} \binom{k}{r} d_r(t). \end{aligned}$$

We also have $d_0(t) = 1$ and $d_1(t) = t$. □

Now that we have Theorem 4.6, we can easily prove the following well-known theorem for the Bell numbers.

Theorem 4.7. A recurrence relation for the Bell numbers is given by $B_{k+1} = \sum_{r=0}^{k} \binom{k}{r} B_r$, where $B_0 = 1$.

Proof. Since $d_k(t) = \sum_{r=1}^{k} \left\{ {k \atop r} \right\} t^r$, we have $d_k(1) = \sum_{r=1}^{k} \left\{ {k \atop r} \right\} = B_k$. And since $d_{k+1}(t) = t \cdot \sum_{r=0}^{k} \binom{k}{r} d_r(t)$, we have $d_{k+1}(1) = \sum_{r=0}^{k} \binom{k}{r} d_r(1)$ or $B_{k+1} = \binom{k}{0} d_0(1) + \sum_{r=1}^{k} \binom{k}{r} B_r = \sum_{r=0}^{k} \binom{k}{r} B_r$ if we define $d_0(1) = B_0 = 1$. □

Theorem 4.2 is a useful, but special kind of recurrence formula for the object distribution polynomials $g_n(t)$. Theorems 4.11 and 4.16 provide more general recurrence formulas that have much broader applications.

Objects of Arbitrary Specification in Identical Boxes

Theorem 4.8. Let $g_m(t)$ be the object distribution polynomial for distributing objects of specification m into identical boxes with no box empty, where $m > 1$. Then

$$\frac{d}{dt} g_m(t) = \sum_{\substack{d^k \mid m \\ d > 1}} g_{\frac{m}{d^k}}(t) t^{k-1}$$

$$= \sum_{\substack{d \mid n \\ d > 1}} \sum_{k \geq 1} g_{\frac{m}{d^k}}(t) t^{k-1}.$$

Proof. The logarithmic generating function for the object distribution polynomials is given by Theorem 2.8:

$$G(x,t) = \sum_{n=1}^{\infty} g_n(t) x^{\log n} = \prod_{d > 1} \frac{1}{(1 - tx^{\log d})} \tag{4.5}$$

Taking the natural logarithm of (4.5), we obtain

$$\log G(x,t) = -\sum_{d > 1} \log(1 - tx^{\log d}). \tag{4.6}$$

Differentiating (4.6) with respect to t gives

$$\frac{\frac{d}{dt} G(x,t)}{G(x,t)} = \sum_{d > 1} \frac{x^{\log d}}{(1 - tx^{\log d})},$$

or, using (4.5), simply $\sum_{n=1}^{\infty} \frac{d}{dt} g_n(t) x^{\log n} = \sum_{n=1}^{\infty} g_n(t) x^{\log n} \cdot \sum_{d > 1} \frac{x^{\log d}}{(1 - tx^{\log d})}$. Now, to obtain the result in the theorem, all we have to do is perform a *tour de force* of algebraic manipulation as follows:

$$\sum_{n=1}^{\infty} \frac{d}{dt} g_n(t) x^{\log n} = \sum_{n=1}^{\infty} g_n(t) x^{\log n} \cdot \sum_{d>1} \frac{x^{\log d}}{(1-tx^{\log d})}.$$

$$= \sum_{n=1}^{\infty} \sum_{d>1} g_n(t) \frac{x^{\log d} x^{\log n}}{(1-tx^{\log d})}$$

$$= \sum_{n=1}^{\infty} \sum_{d>1} g_n(t) x^{\log(dn)} (1 + tx^{\log d} + t^2 x^{\log d^2} + t^3 x^{\log d^3} + \ldots)$$

$$= \sum_{n=1}^{\infty} \sum_{d>1} g_n(t) (x^{\log(dn)} + tx^{\log(d^2 n)} + t^2 x^{\log(d^3 n)} + t^3 x^{\log(d^4 n)} + \ldots)$$

$$= \sum_{n=1}^{\infty} \sum_{d>1} \sum_{k=1}^{\infty} g_n(t) t^{k-1} x^{\log(d^k n)}$$

Now let $m = d^k n$. We then get

$$\sum_{n=1}^{\infty} \frac{d}{dt} g_n(t) x^{\log n} = \sum_{m=2}^{\infty} \sum_{d>1} \sum_{k=1}^{\infty} g_{\frac{m}{d^k}}(t) t^{k-1} x^{\log m}$$

$$= \sum_{m=2}^{\infty} \left(\sum_{\substack{d^k | m \\ d>1}} g_{\frac{m}{d^k}}(t) t^{k-1} \right) x^{\log m}.$$

We are almost finished, but we need to employ a couple more "tricks." First, note that the variable n on the left-hand side of the previous equation is a "dummy variable." So, we can replace n by m on the left-hand side. Second, note that if $m = 1$, then $g_1(t) = 1$. Then $\frac{d}{dt} g_1(t) = \frac{d}{dt} 1 = 0$. Therefore, we have

$\sum_{m=1}^{\infty} \frac{d}{dt} g_m(t) x^{\log m} = \sum_{m=2}^{\infty} \frac{d}{dt} g_m(t) x^{\log m}$. Finally, we have the following result:

$$\sum_{m=2}^{\infty} \frac{d}{dt} g_m(t) x^{\log m} = \sum_{m=2}^{\infty} \left(\sum_{\substack{d^k | m \\ d>1}} g_{\frac{m}{d^k}}(t) t^{k-1} \right) x^{\log m}. \tag{4.7}$$

Equating the coefficients of $x^{\log m}$ on both sides of equation (4.7) immediately gives the desired result stated in the theorem. □

Example 4.9. Let $m = 4$. The divisors d^k of m, written as $d^k | m$, are: $2^1 | 4$, $2^2 | 4$, and $4^1 | 4$. Note that we must take into account *all* divisors of the form d^k that divide m (with $d > 1$). We know that $g_4(t) = t + t^2$. So we have $\frac{d}{dt} g_4(t) = 1 + 2t$. Using Theorem 4.8, we should get the same answer:

$$\sum_{\substack{d^k|4 \\ d>1}} g_{\frac{4}{d^k}}(t) t^{k-1} = g_{\frac{4}{2^1}}(t) t^{1-1} + g_{\frac{4}{2^2}}(t) t^{2-1} + g_{\frac{4}{4^1}}(t) t^{1-1}$$

$$= g_2(t) t^0 + g_1(t) t^1 + g_1(t) t^0$$

$$= g_2(t) + t g_1(t) + g_1(t)$$

$$= t + t \cdot 1 + 1$$

$$= 1 + 2t.$$

What does Theorem 4.8 tell us? Essentially, Theorem 4.8 tells us that the derivative of an object distribution polynomial, $g_m(t)$, is a weighted sum of other object distribution polynomials, where the "weights" are powers of t (i.e., t^{k-1}). We can use Theorem 4.8 to establish a lower bound on the instantaneous slope of an object distribution polynomial at $t = 1$.

Corollary 4.10. *The instantaneous slope of the object distribution polynomial $g_m(t)$ at the point $t = 1$ is at least $\tau(m) - 1$, where $\tau(m)$ is the total number of positive integer divisors of the object specification number m:*

$$\frac{d}{dt} g_m(t) \Big|_{t=1} \geq \tau(m) - 1.$$

Proof. From Theorem 4.8, we have

$$\frac{d}{dt} g_m(t) = \sum_{\substack{d^k|m \\ d>1}} g_{\frac{m}{d^k}}(t) t^{k-1}$$

$$= \sum_{\substack{d|m \\ d>1}} \sum_{k \geq 1} g_{\frac{m}{d^k}}(t) t^{k-1}$$

$$\geq \sum_{\substack{d|m \\ d>1}} g_{\frac{m}{d}}(t) t^0.$$

(Note that, by definition, distribution polynomials are nonnegative and are undefined for $t < 0$.) Then, for $t = 1$, we have

$$\frac{d}{dt}g_m(t)\Big|_{t=1} \geq \sum_{\substack{d|m \\ d>1}} g_{\frac{m}{d}}(1)$$

$$= \sum_{\substack{d|m \\ d>1}} g(m/d)$$

$$\geq \sum_{\substack{d|m \\ d>1}} g(1)$$

$$= \sum_{\substack{d|m \\ d>1}} 1$$

$$= \tau(m) - 1. \quad \square$$

Theorem 4.11 (Integral Recurrence Formula). *Let $g_m(t)$ be the object distribution polynomial for distributing objects of specification m into identical boxes with no box empty. Then*

$$g_m(t) = \sum_{\substack{d^k|m \\ d>1}} \int_0^t g_{\frac{m}{d^k}}(\tau)\tau^{k-1}d\tau.$$

Proof. Integrating the result of Theorem 4.8, $\frac{d}{dt}g_m(t) = \sum_{\substack{d^k|m \\ d>1}} g_{\frac{m}{d^k}}(t)t^{k-1}$, with respect to the variable t,

gives $g_m(t) = \sum_{\substack{d^k|m \\ d>1}} \int_0^t g_{\frac{m}{d^k}}(\tau)\tau^{k-1}d\tau + C$. We must now determine the constant of integration, C.

However, this is easy. We know that $C = 0$, because if $g_m(t)$ is a valid distribution polynomial, then we cannot put objects into zero boxes. Therefore, the coefficient of t^0 in $g_m(t)$ must be zero. $\square$

Example 4.12. Let $m = 8$. The divisors d^k of m, written as $d^k \mid m$, are: $2^1 \mid 8$, $2^2 \mid 8$, $2^3 \mid 8$, $4^1 \mid 8$, and $8^1 \mid 8$. Note that we must take into account *all* divisors of the form d^k that divide m (with $d > 1$). We can now calculate the distribution polynomial $g_8(t)$ using Theorem 4.11:

$$g_8(t) = \int_0^t g_{\frac{8}{2^1}}(\tau)\tau^{1-1}d\tau + \int_0^t g_{\frac{8}{2^2}}(\tau)\tau^{2-1}d\tau + \int_0^t g_{\frac{8}{2^3}}(\tau)\tau^{3-1}d\tau + \int_0^t g_{\frac{8}{4^1}}(\tau)\tau^{1-1}d\tau + \int_0^t g_{\frac{8}{8^1}}(\tau)\tau^{1-1}d\tau$$

$$= \int_0^t g_4(\tau)d\tau + \int_0^t g_2(\tau)\tau d\tau + \int_0^t g_1(\tau)\tau^2 d\tau + \int_0^t g_2(\tau)d\tau + \int_0^t g_1(\tau)d\tau$$

$$= \int_0^t (\tau + \tau^2)d\tau + \int_0^t \tau^2 d\tau + \int_0^t \tau^2 d\tau + \int_0^t \tau d\tau + \int_0^t 1 d\tau$$

$$= \frac{t^2}{2} + \frac{t^3}{3} + \frac{t^3}{3} + \frac{t^3}{3} + \frac{t^2}{2} + t$$

$$= t + t^2 + t^3.$$

The interpretation of $g_8(t)$ is quite simple. Since $8 = 2^3$, we can let the prime number 2 represent, say, a "red" colored ball. So we have three red balls. The distribution polynomial $g_8(t) = 1t + 1t^2 + 1t^3$ tells us that we can distribute three identical red balls into one box in one way, into two identical boxes in one way, and into three identical boxes in one way. No box is allowed to be empty. Also, the order of the boxes is immaterial. The coefficient of t^k counts the number of distributions into precisely k identical boxes with no box empty.

Theorem 4.13 (Average number of boxes per distribution). *For k distinct objects, the average number of boxes per distribution, taken over all possible distributions of the objects into identical boxes with no box empty, is given by*

$$Avg\left\{\frac{boxes}{distribution}(k)\right\} = \frac{B(k+1) - B(k)}{B(k)} = \frac{\Delta B(k)}{B(k)}.$$

$B(k)$ *is the k-th Bell number.*

Proof. If the object distribution polynomial for the given specification number $m = p_1 p_2 \cdots p_k$ is given by $g_m(t) = t + a_2 t^2 + a_3 t^3 + \ldots + a_r t^r + \ldots + t^k$, then

$$\text{Avg}\left\{\frac{boxes}{distribution}(k)\right\} = \frac{\text{total number of boxes for all distributions}}{\text{total number of distributions}}$$

$$= \frac{1 + 2a_2 + 3a_3 + \ldots + ra_r + \ldots + k}{1 + a_2 + a_3 + \ldots + a_r + \ldots + 1}$$

$$= \frac{\frac{d}{dt}g_m(t)}{g_m(t)}\bigg|_{t=1}$$

$$= \frac{\sum_{j=1}^{k} j \left\{{k \atop j}\right\}}{\sum_{j=1}^{k} \left\{{k \atop j}\right\}}$$

$$= \frac{B(k+1) - B(k)}{B(k)}.$$

Using the recurrence relation presented following Defintion 4.4, we have $\left\{{k+j \atop j}\right\} - \left\{{k \atop j-1}\right\} = j\left\{{k \atop j}\right\}$. Summing over all j gives $B(k+1) - B(k) = \sum_j j\left\{{k \atop j}\right\}$. □

Theorem 4.11 provides us with a recurrence relation for calculating the object distribution polynomials $g_n(t)$ that count the number of ways to distribute objects of specification n into identical boxes with no box empty. Now we will derive another form of Theorem 4.11 that is sometimes easier to use in practice.

Definition 4.14 (Integer Power Polynomial). Let $P_n(t) = \sum_{k=1}^{\infty} N(n, k)t^k$, where $N(n, k) = 1$ if the positive integer n can be represented as the kth power of a positive integer, and $N(n, k) = 0$ otherwise.

Example 4.15. Let $n = 16$. Then $P_{16}(t) = t^1 + t^2 + t^4$ because $16 = 16^1$, $16 = 4^2$, and $16 = 2^4$. Clearly, $P_1(t) = t + t^2 + t^3 + \ldots = \frac{t}{1-t}$. Note that many positive integers, m, cannot be represented as a power in more than one way, and in those cases $P_m(t) = t$.

Theorem 4.16 (Second Integral Recurrence Formula). *Let $g_n(t)$ be the object distribution polynomial that counts the number of ways to distribute objects of specification n into identical boxes with no box empty. Then*

$$g_n(t) = \sum_{\substack{d|n \\ d \neq n}} \int_0^t \frac{1}{u} g_d(u) P_{\frac{n}{d}}(u) \, du$$

Note that $n \neq 1$ since $d \neq n$ in the hypothesis.

Proof. We start with the generating function for the distribution of objects into identical boxes with no box empty.

$$\prod_{d>1} \frac{1}{(1-tx^{\log d})} = \sum_{n=1}^{\infty} g_n(t) x^{\log n}$$

$$\log\left(\prod_{d>1}(1-tx^{\log d})^{-1}\right) = \log\left(\sum_{n=1}^{\infty} g_n(t) x^{\log n}\right)$$

$$\sum_{d=2}^{\infty} \left(\log(1-tx^{\log d})^{-1}\right) = \log\left(\sum_{n=1}^{\infty} g_n(t) x^{\log n}\right)$$

$$-\sum_{d=2}^{\infty} \log(1-tx^{\log d}) = \log\left(\sum_{n=1}^{\infty} g_n(t) x^{\log n}\right)$$

Next, we take the derivative with respect to the variable t:

$$-\sum_{d=2}^{\infty} -x^{\log d}\left(\frac{1}{(1-tx^{\log d})}\right) = \left(\sum_{n=1}^{\infty} \frac{d}{dt} g_n(t) x^{\log n}\right)\left(\frac{1}{\sum_{n=1}^{\infty} g_n(t) x^{\log n}}\right)$$

$$\left(\sum_{n=1}^{\infty} g_n(t) x^{\log n}\right)\left(\sum_{d=2}^{\infty} \frac{x^{\log d}}{(1-tx^{\log d})}\right) = \sum_{n=1}^{\infty} \frac{d}{dt} g_n(t) x^{\log n} \tag{4.8}$$

But what is $\sum_{d=2}^{\infty} \frac{x^{\log d}}{(1-tx^{\log d})}$?

First, note that $t\sum_{d=2}^{\infty} \frac{x^{\log d}}{(1-tx^{\log d})} = \sum_{d=2}^{\infty}(tx^{\log d} + t^2 x^{2\log d} + t^3 x^{3\log d} + \ldots)$. When we expand this sum and collect like terms we find that for each $x^{\log n}$ we will get a contribution to $x^{\log n}$ each time n has the form d^k for some positive integer d. The coefficient of $x^{\log 16}$, for example, is $t + t^2 + t^4$ since

$(t+t^2+t^4)x^{\log 16} = tx^{\log 16} + t^2 x^{\log 16} + t^4 x^{\log 16} = tx^{\log 16^1} + t^2 x^{\log 4^2} + t^4 x^{\log 2^4} = tx^{\log 16} + t^2 x^{2\log 4} + t^4 x^{4\log 2}$. Thus, we have

$$t\sum_{d=2}^{\infty} \frac{x^{\log d}}{(1-tx^{\log d})} = \sum_{d=2}^{\infty} P_n(t) x^{\log n} \tag{4.9}$$

Multiplying equation (4.8) by t gives

$$\left(\sum_{n=1}^{\infty} g_n(t) x^{\log n}\right)\left(\sum_{n=2}^{\infty} P_n(t) x^{\log n}\right) = \sum_{n=1}^{\infty} t\frac{d}{dt} g_n(t) x^{\log n} \tag{4.10}$$

By the convolution property for logarithmic generating functions, equation (4.10) implies

$$t\frac{d}{dt} g_n(t) = g_n(t) * P_n(t), \text{ for } n>1 \tag{4.11}$$

Note that must take $n>1$ because the $P_n(t) x^{\log n}$ in equation (4.10) are summed from $n=2$ to infinity. Thus, equation (4.11) gives us

$$t\frac{d}{dt} g_n(t) = \sum_{\substack{d|n \\ d \neq n}} g_d(t) P_{\frac{n}{d}}(t) \tag{4.12}$$

Note that $d \neq n$ because if $d = n$ we would get a term $P_1(t)$ in (4.12), but (4.10) does not have a $P_1(t)$ term. Then, with $d \neq n$, $n \neq 1$, because if $n=1$, then $d=1$ is a divisor of n. But $d \neq n$. Equation (4.12) can be rewritten as $\frac{d}{dt} g_n(t) = \frac{1}{t}\sum_{\substack{d|n \\ d \neq n}} g_d(t) P_{\frac{n}{d}}(t)$. Replacing the variable t by the dummy variable of integration, u, and then integrating both sides of the equation from 0 to t, gives the final result:

$$g_n(t) = \sum_{\substack{d|n \\ d \neq n}} \int_0^t \frac{1}{u} g_d(u) P_{\frac{n}{d}}(u) du. \text{ (Note that } g_n(0) = 0 \text{ for all } g_n(t), n \neq 1.) \quad \square$$

Example 4.17. Suppose that we want to distribute k distinct objects into identical boxes with no box empty. We let the object specification number be $n = p_1 p_2 \ldots p_k$ for distinct primes p_j. Then Theorem 4.16 tells us that $g_n(t) = \sum_{\substack{d|n \\ d \neq n}} \int_0^t \frac{1}{u} g_d(u) P_{\frac{n}{d}}(u) du$. Since n is a product of distinct primes, the integer power polynomial, $P_{\frac{n}{d}}(u)$ satisfies $P_{\frac{n}{d}}(u) = u$, because each divisor, d, of n can be represented as an integer power in only one way. (For example, $d = 2 \cdot 3 = 6$ divides $n = 2 \cdot 3 \cdot 5$, and 6 can be represented as an integer power in only one way, namely $6^1 = 6$.) Therefore, we have

$$g_n(t) = \sum_{\substack{d|n \\ d \neq n}} \int_0^t \frac{1}{u} g_d(u) \cdot u \, du$$

$$= \sum_{\substack{d|n \\ d \neq n}} \int_0^t g_d(u) \, du.$$

This result gives us the following corollary to Theorem 4.16.

Corollary 4.18. *If the object specification number, n, is the product of k distinct primes, say $n = p_1 p_2 \ldots p_k$, then*

$$g_n(t) = \sum_{\substack{d|n \\ d \neq n}} \int_0^t g_d(u) \, du.$$

Corollary 4.19. *Let $B(k)$ denote the kth Bell number. If $n = p_1 p_2 \ldots p_k$, for distinct primes p_j, then*

$$B(k) = \sum_{\substack{d|n \\ d \neq n}} \int_0^1 g_d(u) \, du.$$

Proof. If $n = p_1 p_2 \ldots p_k$ for distinct primes p_j, then $g_n(t)$ is the object distribution polynomial for distributing distinct objects into identical boxes with no box empty. But these distributions are counted by the Stirling numbers of the second kind, $\begin{Bmatrix} k \\ r \end{Bmatrix}$. So, we have $g_n(t) = \sum_{r=1}^{k} \begin{Bmatrix} k \\ r \end{Bmatrix} t^r$. Then the kth Bell number is, by definition, the sum of the Stirling numbers of the second kind. So, we have

$$B(k) = \sum_{r=1}^{k} \begin{Bmatrix} k \\ r \end{Bmatrix}$$

$$= g_n(1)$$

$$= \sum_{\substack{d|n \\ d \neq n}} \int_0^1 g_d(u) \, du \quad \text{for } n = p_1 p_2 \ldots p_k. \quad \square$$

Corollary 4.20. *Let $S(r, t)$ be the rth polynomial for Stirling numbers of the second kind, where*
$$S(r, t) = \sum_{j=1}^{r} \begin{Bmatrix} r \\ j \end{Bmatrix} t^j.$$
Let $S(0, t) = 1$. Then the kth Bell number is given by
$$B(k) = \sum_{r=0}^{k-1} \binom{k}{r} \int_0^1 S(r, t)\,dt.$$

Proof. From Corollary 4.19, we have
$$B(k) = \sum_{\substack{d|n=p_1 p_2 \cdots p_k \\ d \neq n}} \int_0^1 g_d(u)\,du$$
$$= \binom{k}{0}\int_0^1 1\,du + \binom{k}{1}\int_0^1 g_{p_1}(u)\,du + \ldots + \binom{k}{k-1}\int_0^1 g_{p_1 p_2 \cdots p_{k-1}}(u)\,du$$
$$= \sum_{r=0}^{k-1} \binom{k}{r} \int_0^1 S(r, t)\,dt. \quad \square$$

Let us now do an example of how to calculate an object distribution polynomial using Theorem 4.16.

Example 4.21. Let $n = 2^4 = 16$. With the prime being 2, taken to the fourth power, the object distribution polynomial $g_{16}(t)$ will give us the number of ways to distribute 4 identical objects into identical boxes with no box empty. Equivalently, $g_{16}(t)$ gives the integer partitions of the number 4. Using Theorem 4.16, Definition 4.14, and Appendix A, we have

$$g_{16}(t) = \sum_{\substack{d|16 \\ d \neq 16}} \int_0^t \frac{1}{u} g_d(u) P_{\frac{n}{d}}(u)\,du$$
$$= \int_0^t \frac{1}{u}\left(g_1(u)P_{16}(u) + g_2(u)P_8(u) + g_4(u)P_4(u) + g_8(u)P_2(u)\right)du$$
$$= \int_0^t \frac{1}{u}\left(1 \cdot (u + u^2 + u^4) + u(u + u^3) + (u + u^2)^2 + (u + u^2 + u^3)u\right)du$$
$$= \int_0^t (1 + 4u + 3u^2 + 4u^3)\,du$$
$$= t + 2t^2 + t^3 + t^4.$$

You can verify this result in Appendix A, Table 1, for $n = 16 = 2^4 = p_1^4$.

Example 4.22. Let n = $36 = 2^2 \cdot 3^2$. The object distribution polynomial $g_{36}(t)$ will give us the number of ways to distribute, say, two red balls and two blue balls into identical boxes with no box empty.

$$g_{36}(t) = \sum_{\substack{d|36 \\ d \neq 36}} \int_0^t \frac{1}{u} g_d(u) P_{\frac{36}{d}}(u)\, du$$

$$= \int_0^t \frac{1}{u} \left(g_1(u) P_{36}(u) + g_2(u) P_{18}(u) + \ldots + g_{18}(u) P_2(u) \right) du$$

$$= \int_0^t \frac{1}{u} \left(1 \cdot (u + u^2) + u \cdot u + \ldots + (u + 2u^2 + u^3)u \right) du$$

$$= t + 4t^2 + 3t^3 + t^4.$$

Example 4.22 generalizes to the following corollary of Theorem 4.16.

Corollary 4.23. *If* $n = p_1^2 p_2^2 \ldots p_k^2$, *then*

$$g_n(t) = \sum_{d^2|n} \int_0^t g_{d^2}(t)(1+u)\, du + \sum_{\substack{d|n \\ d \neq \text{square}}} \int_0^t g_d(u)\, du.$$

Theorem 4.24. $g(n) \geq \sum_{\substack{d|n \\ d \neq n}} \int_0^1 g_d(t)\, dt$

Proof. By Theorem 4.16, we have

$$g_n(t) = \sum_{\substack{d|n \\ d \neq n}} \int_0^t \frac{1}{u} g_d(u) P_{\frac{n}{d}}(u)\, du$$

$$= \sum_{\substack{d|n \\ d \neq n}} \int_0^t \frac{1}{u} g_d(u)(u + a_2(d)u^2 + a_3(d)u^3 + \ldots)\, du$$

$$= \sum_{\substack{d|n \\ d \neq n}} \int_0^t \frac{1}{u} g_d(u) u\, du + \sum_{\substack{d|n \\ d \neq n}} \int_0^t \frac{1}{u} g_d(u)(a_2(d)u^2 + a_3(d)u^3 + \ldots)\, du$$

In the above, $a_i(d) = 0$ or 1 and depends on d, and $P_{\frac{n}{d}}(u) = (u + a_2(d)u^2 + a_3(d)u^3 + \ldots)$. Then, since $g_n(1) = g(n)$, setting $t = 1$ gives

$$g(n) = \sum_{\substack{d|n \\ d \neq n}} \int_0^1 g_d(u)\,du + \sum_{\substack{d|n \\ d \neq n}} \int_0^1 \frac{1}{u} g_d(u)(a_2(d)u^2 + a_3(d)u^3 + \ldots)\,du, \text{ or } g(n) \geq \sum_{\substack{d|n \\ d \neq n}} \int_0^1 g_d(u)\,du \quad \square$$

Since u is just a dummy variable of integration, we can replace u by t to immediately get the right-hand side of Theorem 4.24.

Example 4.25. Let $n = 16$. Since $g_{16}(t) = t + 2t^2 + t^3 + t^4$, $g_{16}(1) = g(16) = 1 + 2 + 1 + 1 = 5$. Also, the right-hand side of the inequality in Theorem 4.24 gives the following:

$$\sum_{\substack{d|16 \\ d \neq 16}} \int_0^1 g_d(t)\,dt = \int_0^1 (g_1(t) + g_2(t) + g_4(t) + g_8(t))\,dt$$

$$= \int_0^1 (1 + t + (t + t^2) + (t + t^2 + t^3))\,du$$

$$= \int_0^1 (1 + 3t + 2t^2 + t^3)\,du$$

$$= 1 + \tfrac{3}{2} + \tfrac{2}{3} + \tfrac{1}{4}$$

$$= 3.42$$

And $g(16) = 5 \geq 3.42$ as the theorem states.

Theorem 4.24 can be used to obtain a crude lower bound on $g(n)$ as follows. While the lower bound is nothing to boast about, it is interesting because it is expressed in terms of the well-known number-theoretic divisors function, $\tau(n)$.

Corollary 4.26. *Let $\tau(n)$ denote the number of positive integer divisors of a positive integer n. Then* $g(n) \geq \tfrac{1}{2}\tau(n)$.

Proof. If $n = 1$, then $\tau(1) = 1$ and $g(1) = 1$. So, $1 \geq \tfrac{1}{2}$. By Theorem 4.24, if $n > 1$, we have

$$g(n) \geq \sum_{\substack{d|n \\ d \neq n}} \int_0^1 g_d(t)\,dt$$

$$\geq \int_0^1 1\,dt + \sum_{\substack{d|n \\ d \neq 1,n}} \int_0^1 t\,dt$$

$$= 1 + \sum_{\substack{d|n \\ d \neq 1,n}} \tfrac{1}{2}$$

$$= 1 + \tfrac{1}{2} \sum_{\substack{d|n \\ d \neq 1,n}} 1$$

$$= 1 + \tfrac{1}{2}(\tau(n) - 2)$$

$$= \tfrac{1}{2}\tau(n). \quad \square$$

We can use the Integral Recurrence Formula, Theorem 4.11, to prove a couple additional useful recurrence relations.

Theorem 4.27. *Let $g(n)$ denote the total number of ways to distribute objects of specification n into identical boxes with no box empty. Suppose also that there are m objects (primes in n). A recurrence relation for $g(n)$ is given by*

$$g(n) = \sum_{r=0}^m \sum_{\substack{d^k|n \\ d^k \neq 1}} \frac{g(n/d^k, r)}{(k+r)} \tag{4.13}$$

Initial conditions for the recurrence are discussed below.

Proof. By the Integral Recurrence Formula, Theorem 4.11, we have

$$g_n(t) = \sum_{\substack{d^k|n \\ d^k \neq 1}} \int_0^t x^{k-1} g_{\frac{n}{d^k}}(x)\,dx$$

$$= \sum_{\substack{d^k|n \\ d^k \neq 1}} \int_0^t x^{k-1} \left(\sum_{r=0}^m g(n/d^k, r) x^r \right) dx$$

$$= \sum_{\substack{d^k|n \\ d^k \neq 1}} \int_0^t \sum_{r=0}^m g(n/d^k, r) x^{k+r-1}\,dx$$

$$= \sum_{\substack{d^k|n \\ d^k \neq 1}} \sum_{r=0}^{m} g(n/d^k, r) \int_0^t x^{k+r-1}\, dx$$

$$= \sum_{\substack{d^k|n \\ d^k \neq 1}} \sum_{r=0}^{m} g(n/d^k, r) \frac{t^{k+r}}{(k+r)}.$$

Then, since $g(n) = g_n(1)$, we have

$$g(n) = \sum_{r=0}^{m} \sum_{\substack{d^k|n \\ d^k \neq 1}} \frac{g(n/d^k, r)}{(k+r)}. \quad \square$$

Theorem 4.28. *The coefficient of t^r in $g_n(t)$, which is the number of ways to distribute objects of specification n into precisely r identical boxes with no box empty, is given by*

$$g(n, r) = \frac{1}{r} \sum_{\substack{d^k|n \\ d^k \neq 1}} g(n/d^k, r-k) \tag{4.14}$$

Initial conditions for the recurrence are discussed below.

Proof. By the proof of Theorem 4.27, $g_n(t) = \sum_{\substack{d^k|n \\ d^k \neq 1}} \sum_{r=0}^{m} g(n/d^k, r) \frac{t^{k+r}}{(k+r)}$. Also, we have

$g_n(t) = \sum_{r=0}^{m} g(n, r) t^r$. Equating these two expressions for $g_n(t)$, we have

$$\sum_{\substack{d^k|n \\ d^k \neq 1}} \sum_{r=0}^{m} g(n/d^k, r) \frac{t^{k+r}}{(k+r)} = \sum_{r=0}^{m} g(n, r) t^r$$

$$\sum_{r=0}^{m} \left(\sum_{\substack{d|n \\ d \neq 1}} \frac{g(n/d, r)}{(r+1)} t^{r+1} + \sum_{\substack{d^2|n \\ d^2 \neq 1}} \frac{g(n/d^2, r)}{(r+2)} t^{r+2} + \ldots \right) = \sum_{r=0}^{m} g(n, r) t^r$$

Equating the coefficients of t^r on both sides of the last equation gives

$$g(n, r) = \sum_{\substack{d|n \\ d \neq 1}} \frac{g(n/d, r-1)}{r} + \sum_{\substack{d^2|n \\ d^2 \neq 1}} \frac{g(n/d^2, r-2)}{r} + \ldots$$

$$= \frac{1}{r} \sum_{\substack{d^k|n \\ d^k \neq 1}} g(n/d^k, r-k). \quad \square$$

Initial Conditions. In using Theorems 4.27 and 4.28, we need the proper initial conditions, as follows:

1. $g(n, 0) = \begin{cases} 1, & n = 1 \\ 0, & n > 1 \end{cases}$

2. $g(1, r) = \begin{cases} 1, & r = 0 \\ 0, & r > 0 \end{cases}$

3. $g(n, r) = 0$, for $r < 0$

4. $g(n, 1) = \begin{cases} 1, & n > 1 \\ 0, & n = 1 \end{cases}$

Example 4.29. Suppose that we have two red balls and one blue ball, $S = \{red, red, blue\} = \{2, 2, 3\}$. The specification number is $n = 2 \cdot 2 \cdot 3 = 12$. So, the object distribution polynomial is $g_{12}(t) = t + 2t^2 + t^3$. In this polynomial, the coefficient of t^r is $g(12, r)$. Thus, $g(12, 1) = 1$, $g(12, 2) = 2$, and $g(12, 3) = 1$. Let's verify that $g(12, 2) = 2$ using Theorem 4.28 and the initial conditions specified above. The divisors of the 12 having the form d^k, for $d^k \neq 1$, are $\{2^1, 2^2, 3^1, 4^1, 6^1, 12^1\}$. Note that we must include *both* 2^2 and 4^1. Then we have

$$g(12, 2) = \frac{1}{2} \sum_{\substack{d^k \mid 12 \\ d^k \neq 1}} g(12/d^k, 2-k)$$

$$= \frac{1}{2}\begin{pmatrix} g(12/2^1, 2-1) + g(12/3^1, 2-1) + g(12/4^1, 2-1) + g(12/6^1, 2-1) \\ + g(12/12^1, 2-1) + g(12/2^2, 2-2) \end{pmatrix}$$

$$= \frac{1}{2}\big(g(6, 1) + g(4, 1) + g(3, 1) + g(2, 1) + g(1, 1) + g(3, 0)\big)$$

$$= \frac{1}{2}(1+1+1+1+0+0)$$

$$= 2.$$

The recurrence relations presented in Theorems 4.27 and 4.28 are not practical for manual calculations. They are most useful if we write a computer program.

Corollary 4.26 gives us a lower bound on the discrete counting function $g(n)$. We can also use Corollary 4.26 to prove other interesting results in number theory and combinatorics. Here is one example:

Theorem 4.30. *Let $\tau(n)$ denote the total number of positive integer divisors of n. The infinite series* $\sum_{n=1}^{\infty} \frac{\tau(n)}{n^2}$ *converges. Furthermore,* $\sum_{n=1}^{\infty} \frac{\tau(n)}{n^2} \leq 4$.

Proof. By Corollary 4.26, $g(n) \geq \frac{1}{2}\tau(n)$. In Theorem 3.5 we proved that $\sum_{n=1}^{\infty} \frac{g(n)}{n^2} = 2$. Putting these facts together, we have

$$\tau(n) \leq 2g(n)$$

$$\frac{\tau(n)}{n^2} \leq \frac{2g(n)}{n^2}$$

$$\sum_{n=1}^{\infty} \frac{\tau(n)}{n^2} \leq 2\sum_{n=1}^{\infty} \frac{g(n)}{n^2}$$

$$\sum_{n=1}^{\infty} \frac{\tau(n)}{n^2} \leq 4 \quad \square$$

The upper bound in Theorem 4.30 is not a bad result. For example, note that, using empirical data, we have $\sum_{n=1}^{60} \frac{\tau(n)}{n^2} = 2.60325$.

Exercises

1. Prove that if the positive integer n is composite, then $\frac{d}{dt} g_n(t) \geq \frac{1}{2}\tau(n)$ at $t=1$.

2. Let $n = p_1^2 p_2^2 q$ for distinct primes p_1, p_2, q. Use Theorem 4.2 and Appendix A to calculate $g_n(t)$.

3. Use the recurrence relation in Theorem 4.7 to calculate the Bell numbers, B_n, for $n = 1, 2, 3, 4, 5$.

4. For five distinct objects, what is the average number of boxes per distribution when we distribute the objects into identical boxes with no box empty in all possible ways?

5. Use Theorem 4.16 to calculate the object distribution polynomial $g_{18}(t)$.

6. What is the combinatorial meaning of $g_{18}(t)$? Why is $g_{12}(t) = g_{18}(t)$?

Chapter 5. Distributions in Identical Boxes with Distinct Occupancies

In how many different ways can we distribute objects of arbitrary specification, such as colored balls, into identical boxes such that the occupancies of the boxes are distinct? We restrict the problem to the requirement that no box can be empty. (The case where empty boxes are allowed is a trivial consequence of our present problem, obtained by simple summation over the number of boxes.) Note that the order of the balls within a box does not matter. Within a given box, we can rearrange the order of the balls without changing its occupancy. Also, since the boxes themselves are identical, the order of the boxes does not matter, because we cannot tell the empty boxes apart. Only when we place objects into the boxes do they acquire an "identity."

By "distinct occupancy," we mean that no two boxes contain the same distribution of colored balls. Suppose, for example, that we have two red balls and two blue balls. There are clearly many different ways that we can distribute the balls into identical boxes with no box empty. If we distribute the balls into two boxes, then some of the distributions will have distinct occupancy, and some will not. Figure 5.1 shows an example for each kind of occupancy.

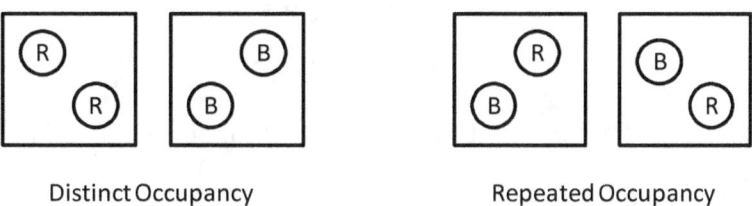

Distinct Occupancy   Repeated Occupancy

Figure 5.1. Kinds of occupancy: distinct and repeated (non-distinct).

To calculate the number of ways to distribute colored balls of specification n into identical boxes with distinct occupancy (i.e., no two boxes contain the same multiset of objects), we can build a restricted form of the logarithmic generating function:

Theorem 5.1. *The logarithmic generating function for the distribution of colored balls of specification n into identical boxes with distinct occupancies, and no box empty, is given by*

$$\Delta(x,t) = \prod_{d>1}(1+tx^{\log d}) = \sum_{n=1}^{\infty}\delta_n(t)x^{\log n}.$$

Proof. From Theorem 2.8, we start with the general form of the logarithmic generating function for arbitrary (unrestricted) distributions:

$$G(x,t) = \prod_{d>1}\frac{1}{(1-tx^{\log d})} = \sum_{n=1}^{\infty}g_n(t)x^{\log n}.$$

To prevent the occurrence of identical occupancies in two or more boxes, we can introduce a binary function, ε_1^k, which takes on the value zero if repeated occupancies occur:

$$\varepsilon_1^k \equiv \begin{cases} 1, & \text{if } k=1 \\ 0, & \text{if } k>1 \end{cases}$$

Here, we define $\varepsilon_1 = \varepsilon_1^1 = 1$. The "$k$" is interpreted as the k-th power of ε_1, if k is a positive integer. Then $\Delta(x,t) = G(x,t\varepsilon_1)$ since

$$\begin{aligned} G(x,t\varepsilon_1) &= \prod_{d>1}\frac{1}{(1-t\varepsilon_1 x^{\log d})} \\ &= \prod_{d>1}(1+t\varepsilon_1^1 x^{\log d} + t^2\varepsilon_1^2 x^{2\log d} + t^3\varepsilon_1^3 x^{3\log d} + \ldots) \\ &= \prod_{d>1}(1+tx^{\log d}), \text{ by the properties of } \varepsilon_1^k \\ &= \Delta(x,t) \quad \square \end{aligned}$$

Remark 5.2. Note, for example, that the term $t^2 x^{2\log d} = t^2 x^{\log d^2}$ represents, say, two red balls in two non-empty boxes, or, equivalently, one red ball in each box. This situation is clearly *not* a distinct occupancy distribution. So, we must force the term to equal zero. That is accomplished by ε_1^2, which equals zero by definition. A similar argument applies to other terms (except 1, which is excluded).

Remark 5.3. The coefficient of $x^{\log n}$ is a polynomial $\delta_n(t)$ that solves the distribution problem in question. Specifically, the coefficient of t^k in the polynomial $\delta_n(t)$ is the number of ways to distribute colored balls of specification n into precisely k identical boxes with distinct occupancies (and no box empty).

Example 5.4. Suppose that we have two red balls and one blue ball. Let $p_1 = 2 = red$ and $p_2 = 3 = blue$. The specification number is $n = p_1^2 p_2 = 2^2 \cdot 3 = 12$. In how many ways can we distribute the three balls into identical boxes with distinct occupancies and no box empty? By Theorem 5.1, we must expand the generating function $\Delta(x,t) = (1 + tx^{\log 2})(1 + tx^{\log 3})(1 + tx^{\log 4})\ldots$. This is clearly a product involving infinitely many factors. However, since we are only interested in the coefficient of $x^{\log 12}$, we can ignore all terms of the form $(1 + tx^{\log d})$ for $d > 12$. Also, the only terms that contribute to $x^{\log 12}$ are those terms where d is a divisor of 12. Thus, we need only consider the product:

$$(1+tx^{\log 2})(1+tx^{\log 3})(1+tx^{\log 4})(1+tx^{\log 6})(1+tx^{\log 12}) = 1 + tx^{\log 2} + tx^{\log 3} + tx^{\log 4} + (t+t^2)x^{\log 6} + \ldots + (t+2t^2)x^{\log 12}.$$

Since the coefficient of $x^{\log 12}$ is $t + 2t^2$, we have $\delta_{12}(t) = t + 2t^2$. The coefficient of t is 1, so there is only one way to distribute two red balls and one blue ball into one box with distinct occupancy:

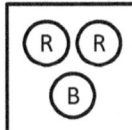

The coefficient of t^2 is 2, so there are two ways to distribute two red balls and one blue ball into two identical boxes with distinct occupancies:

Note that if two boxes have the same occupancy (i.e., contain the same specification of colored balls), then the colored balls in the two boxes can be paired by like colors. Thus, in order to have boxes with repeated, or non-distinct occupancies, it is necessary to have at least two balls of the same color. Equivalently, if the specification n of colored balls consists entirely of distinct colors (i.e., precisely one ball of each color), then *all* of the distributions will be distinct. Therefore, we have the following lemma:

Lemma 5.5. *Let $n = p_1 p_2 ... p_k$ for distinct primes p_j. Then $\delta_n(t) = g_n(t)$, where $\delta_n(t)$ is the distribution polynomial for distinct occupancy, and $g_n(t)$ is the distribution polynomial for arbitrary, or general occupancy.*

A Recurrence Relation

In a certain sense, we could argue that the distribution problem for distinct occupancy is completely solved. The entire solution is contained in Theorem 5.1. Unfortunately, as we have seen in Example 5.4, expanding the generating function, by multiplying its terms, involves a great many multiplications. Without simplifications of any kind, evaluating the coefficient of $x^{\log 12}$ would require $2^{11} = 2048$ multiplications. (The exponent is 11, instead of 12, because we exclude 1 as a divisor of 12.) We did, of course, make a few simplifications in Example 5.4. What we really need, for the sake of computational efficiency, is a *recurrence relation*. The recurrence relation will allow us to calculate distribution polynomials $\delta_n(t)$ in terms of other distribution polynomials for smaller values of n. Theorems 5.6 and 5.8 are central to this effort[1].

Theorem 5.6. *A recurrence relation for* $\delta_n(t)$ *is given by*

$$\delta_n(t) \cdot \log n = \sum_{d>1} \sum_{j\geq 1} \delta_{\frac{n}{d^j}}(t) \cdot (-1)^{j+1} \cdot t^j \cdot \log d$$

where d^j is a divisor of n and $\delta_1(t) = 1$.

Proof. By Theorem 5.1, $\Delta(x,t) = \prod_{d>1}(1+tx^{\log d}) = \sum_{n=1}^{\infty} \delta_n(t) x^{\log n}$. Taking the natural logarithm of both sides of the equation (in reverse order) gives

$$\log(\sum_{n=1}^{\infty} \delta_n(t) x^{\log n}) = \log(\prod_{d=2}^{\infty}(1+tx^{\log d}))$$

$$= \sum_{d=2}^{\infty} \log(1+tx^{\log d})$$

Taking the derivative of both sides of the equation with respect to x gives

$$\frac{d}{dx}(\log(\sum_{n=1}^{\infty} \delta_n(t) x^{\log n})) = \frac{d}{dx}(\sum_{d=2}^{\infty} \log(1+tx^{\log d}))$$

$$\frac{d}{dx}(\sum_{n=1}^{\infty} \delta_n(t) x^{\log n})(\frac{1}{\sum_{n=1}^{\infty} \delta_n(t) x^{\log n}}) = \frac{d}{dx}(\sum_{d=2}^{\infty} \log(1+tx^{\log d}))$$

$$\frac{\sum_{n=1}^{\infty} \delta_n(t) \log n \cdot x^{\log n - 1}}{\sum_{n=1}^{\infty} \delta_n(t) x^{\log n}} = \sum_{d=2}^{\infty} t \cdot \log d \cdot x^{\log d - 1} \cdot (1+tx^{\log d})^{-1}$$

$$\frac{\frac{1}{x} \cdot \sum_{n=1}^{\infty} \delta_n(t) \log n \cdot x^{\log n}}{\sum_{n=1}^{\infty} \delta_n(t) x^{\log n}} = \frac{1}{x} \cdot \sum_{d=2}^{\infty} t \cdot \log d \cdot x^{\log d} \cdot (1+tx^{\log d})^{-1}$$

---

[1] In some equations, we insert a "dot," for ordinary multiplication to enhance the clarity of the equation. This is typically done where the juxtaposition of alphabetical letters may lead to confusion.

Multiplying through by x and rearranging, we get

$$\sum_{n=1}^{\infty} \delta_n(t) \cdot \log n \cdot x^{\log n} = (\sum_{n=1}^{\infty} \delta_n(t) x^{\log n})(\sum_{d=2}^{\infty} t \cdot \log d \cdot x^{\log d} \cdot (1 + tx^{\log d})^{-1})$$

$$= \sum_{n=1}^{\infty} \sum_{d=2}^{\infty} \delta_n(t) x^{\log n} \cdot t \cdot \log d \cdot x^{\log d} \cdot (1 + tx^{\log d})^{-1}$$

$$= \sum_{n=1}^{\infty} \sum_{d=2}^{\infty} \delta_n(t) x^{\log n} \cdot t \cdot \log d \cdot x^{\log d} \cdot \sum_{k=0}^{\infty} (-1)^k t^k x^{k \log d}$$

$$= \sum_{n=1}^{\infty} \sum_{d=2}^{\infty} \sum_{k=0}^{\infty} \delta_n(t) x^{\log n} \cdot t \cdot \log d \cdot x^{\log d} \cdot (-1)^k t^k x^{k \log d}$$

$$= \sum_{n=1}^{\infty} \sum_{d=2}^{\infty} \sum_{k=0}^{\infty} \delta_n(t) \cdot t^{k+1} \cdot \log d \cdot (-1)^k x^{\log n + \log d + \log d^k}$$

$$= \sum_{n=1}^{\infty} \sum_{d=2}^{\infty} \sum_{k=0}^{\infty} \delta_n(t) \cdot t^{k+1} \cdot \log d \cdot (-1)^k x^{\log(nd^{k+1})}$$

Now let $m = n \cdot d^{k+1}$, which gives

$$\sum_{n=1}^{\infty} \delta_n(t) \cdot \log n \cdot x^{\log n} = \sum_{m=2}^{\infty} \sum_{d=2}^{\infty} \sum_{k=0}^{\infty} \delta_{\frac{m}{d^{k+1}}}(t) \cdot t^{k+1} \cdot (-1)^k \cdot \log d \cdot x^{\log m}.$$

It is now to be understood that d^{k+1} is a divisor of m. Note that as k ranges from 0 to ∞, as d ranges from 2 to ∞, and as n ranges from 1 to ∞, m ranges from 2 to ∞. Since n is a dummy variable of summation on the left-hand-side of the last equation, we can replace n by m in the left-hand summation. Also, if $m = 1$, then $\log m = \log 1 = 0$. So, we have

$$\sum_{m=2}^{\infty} \delta_m(t) \cdot \log m \cdot x^{\log m} = \sum_{m=2}^{\infty} \sum_{d=2}^{\infty} \sum_{k=0}^{\infty} \delta_{\frac{m}{d^{k+1}}}(t) \cdot t^{k+1} \cdot (-1)^k \cdot \log d \cdot x^{\log m}$$

$$\sum_{m=2}^{\infty} (\delta_m(t) \log m) x^{\log m} = \sum_{m=2}^{\infty} (\sum_{d=2}^{\infty} \sum_{k=0}^{\infty} \delta_{\frac{m}{d^{k+1}}}(t) \cdot t^{k+1} \cdot (-1)^k \cdot \log d) x^{\log m}$$

Equating the coefficients of $x^{\log m}$ on both sides of the last equation gives

$\delta_m(t) \log m = \sum_{d=2}^{\infty} \sum_{k=0}^{\infty} \delta_{\frac{m}{d^{k+1}}}(t) \cdot t^{k+1} \cdot (-1)^k \cdot \log d$. Next, we change variables and let $j = k + 1$ to get

$\delta_m(t) \log m = \sum_{d>1} \sum_{j \geq 1} \delta_{\frac{m}{d^j}}(t) \cdot t^j \cdot (-1)^{j-1} \cdot \log d$. Since $(-1)^2 = 1$, we note that

$(-1)^{j-1} \cdot 1 = (-1)^{j-1} \cdot (-1)^2 = (-1)^{j+1}$. This just gives us a slightly nicer way to write the final result:

$\delta_m(t)\log m = \sum_{d>1}\sum_{j\geq 1} \delta_{\frac{m}{d^j}}(t)\cdot t^j \cdot (-1)^{j+1}\cdot \log d$, where d^j is a divisor of m and $\delta_1(t)=1$. Changing variables from m to n gives the form in the theorem statement. □

Example 5.7. Suppose that we have two red balls and two blue balls: $S = \{red, red, blue, blue\}$. Let the prime number 2 represent the color *red*, and let the prime number 3 represent the color *blue*. The specification number is $n = 2^2\cdot 3^2 = 36$. In how many ways can we distribute the colored balls into identical boxes such that each box has a distinct occupancy of colored balls? We also require that no box is empty. To solve this problem, we use the recurrence relation in Theorem 5.6 to calculate the distribution polynomial $\delta_n(t)$. The non-unit divisors d of $n = 36$ are $\{2, 3, 4, 6, 9, 12, 18, 36\}$. Applying Theorem 5.6, we have

$$\delta_{36}(t)\log 36 = \sum_{d>1}\sum_{j\geq 1} \delta_{\frac{36}{d^j}}(t)t^j(-1)^{j+1}\log d$$

$$= \delta_{18}(t)t^1(-1)^2\log 2 + \delta_9(t)t^2(-1)^3\log 2$$
$$+ \delta_{12}(t)t^1(-1)^2\log 3 + \delta_4(t)t^2(-1)^3\log 3$$
$$+ \delta_9(t)t^1(-1)^2\log 4$$
$$+ \delta_6(t)t^1(-1)^2\log 6 + \delta_1(t)t^2(-1)^3\log 6$$
$$+ \delta_4(t)t^1(-1)^2\log 9$$
$$+ \delta_3(t)t^1(-1)^2\log 12$$
$$+ \delta_2(t)t^1(-1)^2\log 18$$
$$+ \delta_1(t)t^1(-1)^2\log 36$$

Note that $\log 36 = \log(2^2\cdot 3^2) = \log 2^2 + \log 3^2 = 2\log 2 + 2\log 3$. Since 2 and 3 are coprime, we can equate coefficients of either $\log 2$ or $\log 3$ on both sides of the equation. This is just an algebraic shortcut. Alternatively, we could equate the coefficients of $\log 36$ on both sides of the equation, but then we must use the rules of logarithms to collect and represent all logarithms in terms of $\log 36$. For this example, let us equate the coefficients of $\log 2$ on both sides of the last equation:

$$2\delta_{36}(t) = t\delta_{18}(t) - t^2\delta_9(t) + 2t\delta_9(t) + t\delta_6(t) - t^2\delta_1(t) + 2t\delta_3(t) + t\delta_2(t) + 2t\delta_1(t) \tag{5.1}$$

Now, if we also happen to know that $\delta_{18}(t) = t + 2t^2$, $\delta_9(t) = t$, $\delta_6(t) = t + t^2$, $\delta_3(t) = \delta_2(t) = t$, and $\delta_1(t) = 1$, then, upon substitution into (5.1) and simplifying, we obtain the final result: $\delta_{36}(t) = t + 3t^2 + t^3$. The coefficient of t^k gives the number of ways to distribute two red balls and two blue balls into precisely k identical boxes, with no box empty, such that each box has a distinct occupancy of colored balls. Figure 5.2 presents all the possible placements corresponding to the distribution polynomial $\delta_{36}(t) = t + 3t^2 + t^3$.

If we did not use the recurrence relation of Theorem 5.6, we would have to perform $2^{35} = 34359738368$ multiplications using Theorem 5.1. Even if we simplified the product by including only the non-unit divisors of 36, we would still require $2^8 = 256$ multiplications in a direct application of Theorem 5.1.

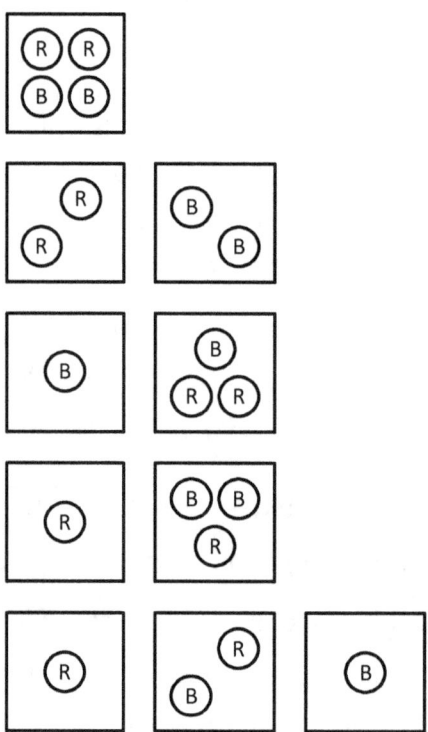

Figure 5.2. Distributions of colored balls corresponding to $\delta_{36}(t) = t + 3t^2 + t^3$.

Table 5.1 provides a short list of several object distribution polynomials for distinct occupancy.

Table 5.1. Distribution polynomials for Distinct Occupancies

Specification form n	Distribution polynomial $\delta_n(t)$
1	1
p_1	t
p_1^2	t
p_1^3	$t + t^2$
$p_1 p_2$	$t + t^2$
$p_1^2 p_2$	$t + 2t^2$
$p_1^2 p_2^2$	$t + 3t^2 + t^3$
$p_1^3 p_2^3$	$t + 7t^2 + 8t^3 + t^4$

Theorem 5.8 (Integral Recurrence Formula). *For distinct occupancy object distribution polynomials, $\delta_n(t)$, we have the following analog of Theorem 4.16:*

$$\delta_n(t) = -\sum_{\substack{d \mid n \\ d \neq n}} \int_0^t \frac{1}{u} \delta_d(u) P_{\frac{n}{d}}(-u) \, du.$$

Proof. Use the generating function $\Delta(x, t) = \prod_{d>1}(1 + tx^{\log d}) = \sum_{n=1} \delta_n(t) x^{\log n}$ and repeat the derivation presented in the proof of Theorem 4.16. □

Identical Objects

Let $n = p^k$ for some prime p and positive integer k. Distributing k identical objects into identical boxes is the same problem as distributing k "units" into identical boxes. Thus, for this special case, $\delta_{p^k}(t)$ gives the number of partitions of k into *distinct parts*. In other words, the coefficient of t^r in $\delta_{p^k}(t)$, denoted by $[t^r]\delta_{p^k}(t)$, is the number of partitions of k into precisely r distinct parts:

$$[t^r]\delta_{p^k}(t) = \pi_d(k,r) \tag{5.2}$$

By Theorem 5.6, $\delta_{p^k}(t) \cdot \log p^k = \sum_{d>1}\sum_{j\geq 1}\delta_{\frac{p^k}{d^j}}(t)\cdot(-1)^{j+1}\cdot t^j \cdot \log d$. Simplifying this expression, we have

$$k \cdot \delta_{p^k}(t)\log p = \sum_{r\geq 1}\sum_{j\geq 1}\delta_{\frac{p^k}{(p^r)^j}}(t)(-1)^{j+1}t^j \log p^r$$

$$= \sum_{r\geq 1}\sum_{j\geq 1}\delta_{\frac{p^k}{(p^r)^j}}(t)(-1)^{j+1}t^j r\log p$$

$$= \sum_{m\geq 1}\sum_{j\geq 1}\delta_{p^{k-m}}(t)(-1)^{j+1}t^j \frac{m}{j}\log p, \text{ where } m = rj$$

$$= \sum_{m=1}^{k}\sum_{j|m}\delta_{p^{k-m}}(t)(-1)^{j+1}t^j \frac{m}{j}\log p$$

$$= \sum_{m=1}^{k}\delta_{p^{k-m}}(t)\sum_{j|m}(-1)^{j+1}\frac{m}{j}t^j \log p$$

Equating coefficients of log p on both sides of the equation, we get

$$k \cdot \delta_{p^k}(t) = \sum_{m=1}^{k}\delta_{p^{k-m}}(t)P_m(t), \text{ where } P_m(t) = \sum_{j|m}(-1)^{j+1}\frac{m}{j}t^j. \tag{5.3}$$

If we now let $t=1$, then $\delta_{p^k}(1)$ gives the total number of partitions of k into distinct parts. So, $\delta_{p^k}(1) = \pi_d(k)$. Let $t = 1$ and $m = k - j$ in (5.3), and, reversing the order of summation, and we obtain the following lemma:

Lemma 5.9. *Let $\pi_d(k)$ denote the number of partitions of a positive integer k into distinct parts. Then*

$$k \cdot \pi_d(k) = \sum_{j=0}^{k-1}\pi_d(j)\beta(k-j),$$

where $\pi_d(0) \equiv 1$ and $\beta(n) = P_n(1) = -\sum_{d|n}(-1)^d \frac{n}{d}$. (We could bring the minus sign into the summation to get $(-1)^{d+1}(n/d)$ as the summand. However, that is less elegant for subtle technical reasons.)

We will return to Lemma 5.9 and prove it after we digress to examine the function $\beta(n)$.

The Sum of Odd Divisors Function

What is the function $\beta(n) = P_n(1) = -\sum_{d|n}(-1)^d \frac{n}{d}$? Calculating the first few values of the function, we obtain $\beta(1) = 1$, $\beta(2) = 1$, $\beta(3) = 4$, $\beta(4) = 1$, $\beta(5) = 6$, $\beta(6) = 4$, ... So, the sequence in question is $\{1, 1, 4, 1, 6, 4, 8, 1, 13, 6, 12, 4, ...\}$. Obviously, the function $\beta(n)$ behaves quite erratically. This is typical of "number-theoretic" functions, and we immediately suspect that $\beta(n)$ might be a function that we already know. Looking up the sequence in Sloane's *Encyclopedia of Integer Sequences* [6], we find that our unknown sequence is listed as number M3197, which is the *sum of odd divisors of n*, or $\sigma_o(n)$. With this insight, we must now prove that $\beta(n) = \sigma_o(n)$. Since $\beta(n)$ has some interesting properties, we will break the proof into three cases. We could combine all three cases into a single proof, but the individual cases are interesting in themselves.

Lemma 5.10. $\beta(n) = \sigma_o(n)$, where $\beta(n) = -\sum_{d|n}(-1)^{n/d} \cdot d$ and $\sigma_o(n)$ is the sum of the odd divisors of n. (Note that $-\sum_{d|n}(-1)^{n/d} d = -\sum_{d|n}(-1)^d \frac{n}{d}$ because if $n = d_1 \cdot d_2$, then $d_1 = \frac{n}{d_2}$, where d_1 and d_2 are both divisors of n.)

Proof.

<u>Case 1.</u> Let $n = 2^k$ for $k = 0, 1, 2, ...$. Then we have

$$\beta(2^k) = -\sum_{d|2^k}(-1)^{2^k/d} \cdot d$$
$$= -(2^0 + 2^1 + ... + 2^{k-1} - 2^k)$$
$$= 2^k - (2^k - 1)$$
$$= 1.$$

Thus, $\beta(2^k) = 1$. Of course, since the only *odd* divisor of 2^k is 1, we also have $\beta(2^k) = \sigma_o(2^k) = 1$. □

<u>Case 2.</u> Let *n* be odd. Then any divisor *d* of *n* is odd, and *n/d* is odd. Then we have

$$\beta(n_{odd}) = -\sum_{d|n_{odd}} (-1)^{odd} \cdot d$$

$$= \sum_{d|n_{odd}} (-1)^{even} \cdot d$$

$$= \sum_{d|n_{odd}} d$$

$$= \sigma(n_{odd})$$

$$= \sigma_o(n_{odd}) \quad \square$$

<u>Case 3</u>. Let $n = 2^k m$ for $k \geq 1$, where m is the "odd part" of n. Then we have

$$\beta(2^k m) = -\sum_{d|2^k m} (-1)^{2^k m/d} \cdot d$$

$$= -((\sum_{d|2^k} (-1)^{2^k m/d} \cdot d) + (\sum_{\substack{d|m \\ d \neq 1}} (-1)^{2^k m/d} \cdot d) + (\sum_{\substack{d|2^k m \\ d \text{ even} \\ \text{odd part}(d) \neq 1}} (-1)^{2^k m/d} \cdot d))$$

$$= -((\sum_{r=0}^{k-1} 2^r - 2^k) + (\sigma(m) - 1) + (\sum_{r=1}^{k-1} 2^r(\sigma(m) - 1) - 2^k(\sigma(m) - 1)))$$

$$= -(\sum_{r=0}^{k-1} 2^r - 2^k + \sum_{r=0}^{k-1} 2^r(\sigma(m) - 1) - 2^k(\sigma(m) - 1))$$

$$= -(2^k - 1 - 2^k + (\sigma(m) - 1)(2^k - 1 - 2^k))$$

$$= \sigma(m)$$

But $\sigma(m)$ is the sum of the odd divisors of $n = 2^k m$, because m is the odd part of n. Therefore, $\beta(2^k m) = \sigma(m) = \sigma_o(2^k m)$. $\square$

Since $\beta(n) = \sigma_o(n)$, we can restate Lemma 5.9 as follows:

Theorem 5.11. *Let $\pi_d(k)$ denote the number of partitions of a positive integer k into distinct parts. Then*

$$k \cdot \pi_d(k) = \sum_{j=0}^{k-1} \pi_d(j) \sigma_o(k - j),$$

where $\pi_d(0) \equiv 1$ and $\sigma_o(n)$ is the sum of odd divisors of n.

Proof. Substitute $\sigma_o(k - j)$ for $\beta(k - j)$ in Lemma 5.9. $\square$

Now that we know that $\sigma_o(n)$ can be represented by the rather peculiar form $\sigma_o(n) = -\sum_{d|n}(-1)^{n/d} \cdot d$, let us take full advantage of this in the next few sections. We can use it to easily prove some interesting theorems.

Application of the Möbius Inversion Formula

If $f(n) = \sum_{d|n} g(d)$, then $g(n) = \sum_{d|n} \mu(d) f(n/d)$, where $\mu(n)$ is the Möbius function defined by

$$\mu(n) = \begin{cases} 1 & \text{if } n = 1 \\ 0 & \text{if } p^2 \mid n \text{ for some prime } p \\ (-1)^r & \text{if } n = p_1 p_2 \ldots p_r \text{ for distinct primes } p_j \end{cases}$$

This is the famous *Möbius inversion formula*. We can apply the Möbius inversion formula to our formula for $\sigma_o(n)$ to obtain the following theorem.

Theorem 5.12. $\sum_{d|n} d \cdot \mu(d) \cdot \sigma_0(n/d) = (-1)^{n+1}$.

Proof. Since $\sigma_o(n) = -\sum_{d|n}(-1)^{n/d} d = -\sum_{d|n}(-1)^d \frac{n}{d}$, we have $-\frac{\sigma_o(n)}{n} = \sum_{d|n} \frac{(-1)^d}{d}$. Applying the Möbius inversion formula, we get $\frac{(-1)^n}{n} = \sum_{d|n} \mu(d)\left(-\frac{\sigma_o(n/d)}{n/d}\right)$. Simplifying the last equation proves the theorem. □

Dirichlet Series Generating Functions

We will now derive a *Dirichlet series generating function* for $\sigma_o(n)$. Let $f(s)$ and $g(s)$ be two number-theoretic functions having the Dirichlet series generating functions $f(s) = \sum_{n=1}^{\infty} \frac{a_n}{n^s}$ and $g(s) = \sum_{n=1}^{\infty} \frac{b_n}{n^s}$. Here, we are not interested in analytic properties such as convergence. We view the generating functions as elements of a formal ring of Dirichlet power series. If we multiply $f(s)$ and $g(s)$, we get

$$f(s)g(s) = \sum_{n=1}^{\infty} \left(\sum_{d|n} a_d b_{\frac{n}{d}} \right) \frac{1}{n^s}.$$

Now we know that $\sigma_o(n) = -\sum_{d|n} (-1)^{n/d} \cdot d$. So, let $a_d = d$ and $b_{\frac{n}{d}} = (-1)^{n/d+1}$. Then the Dirichlet series generating function for $\sigma_o(n)$ is given by

$$\sum_{n=1}^{\infty} \frac{\sigma_o(n)}{n^s} = f(s)g(s) = \left(\sum_{n=1}^{\infty} \frac{n}{n^s} \right) \left(\sum_{n=1}^{\infty} \frac{(-1)^{n+1}}{n^s} \right)$$

$$= \left(\frac{1}{1^s} + \frac{2}{2^s} + \frac{3}{3^s} + \frac{4}{4^s} + \ldots \right) \left(\frac{1}{1^s} - \frac{1}{2^s} + \frac{1}{3^s} - \frac{1}{4^s} + \ldots \right)$$

We could certainly multiply this out, combine like terms, and read off the coefficients to discover, for example, that $\sigma_o(6) = 4$. Instead, let us figure out what generating functions represent $f(s)$ and $g(s)$. Then we can multiply them to get a generating function for $\sigma_o(n)$. Since the *Riemann zeta function* is $\varsigma(s) = \sum_{n=1}^{\infty} \frac{1}{n^s}$, we have

$$f(s) = \sum_{n=1}^{\infty} \frac{n}{n^s} = \sum_{n=1}^{\infty} \frac{1}{n^{s-1}} = \varsigma(s-1)$$

We must now find an expression for $g(s)$.

$$g(s) = \sum_{n=1}^{\infty} \frac{(-1)^{n+1}}{n^s}$$

$$= \frac{1}{1^s} - \frac{1}{2^s} + \frac{1}{3^s} - \frac{1}{4^s} + \frac{1}{5^s} - \frac{1}{6^s} + \ldots$$

$$= (\frac{1}{1^s} + \frac{1}{3^s} + \frac{1}{5^s} + \ldots) - (\frac{1}{2^s} + \frac{1}{4^s} + \frac{1}{6^s} + \ldots)$$

$$= (\frac{1}{1^s} + \frac{1}{3^s} + \frac{1}{5^s} + \ldots) - \frac{1}{2^s}\varsigma(s)$$

$$= (\varsigma(s) - (\frac{1}{2^s} + \frac{1}{4^s} + \frac{1}{6^s} + \ldots)) - \frac{1}{2^s}\varsigma(s)$$

$$= (\varsigma(s) - \frac{1}{2^s}\varsigma(s)) - \frac{1}{2^s}\varsigma(s)$$

$$= \varsigma(s)(1 - \frac{1}{2^{s-1}})$$

Multiplying the generating functions for $f(s)$ and $g(s)$, we obtain the following theorem:

Theorem 5.13. *The Dirichlet series generating function for $\sigma_o(n)$, the sum of odd divisors of n, is given by*

$$\varsigma(s)\varsigma(s-1)(1 - \frac{1}{2^{s-1}}) = \sum_{n=1}^{\infty} \frac{\sigma_o(n)}{n^s}$$

We also have the well-known result for $\sigma(n)$, the sum of divisors of n (Hardy and Wright, Theorem 290, p. 250 [7]):

Theorem 5.14. *The Dirichlet series generating function for $\sigma(n)$, the sum of divisors of n, is given by*

$$\varsigma(s)\varsigma(s-1) = \sum_{n=1}^{\infty} \frac{\sigma(n)}{n^s}$$

Proof. Since $\sigma(n) = \sum_{d|n} d$, let $a_d = d$ and let $b_{\frac{n}{d}} = 1$. Then apply the product rule with

$$f(s) = \varsigma(s-1) = \sum_{n=1}^{\infty} \frac{n}{n^s} \text{ and } g(s) = \varsigma(s) = \sum_{n=1}^{\infty} \frac{1}{n^s}. \quad \square$$

Since $\sigma(n)$ is the sum of divisors of n, and $\sigma_o(n)$ is the sum of *odd* divisors of n, it is natural to define $\sigma_e(n)$ to be the sum of *even* divisors of n. Then, since every divisor of n is either odd or even, we have $\sigma(n) = \sigma_o(n) + \sigma_e(n)$. We can now easily derive a generating function for $\sigma_e(n)$.

Theorem 5.15. *The Dirichlet series generating function for $\sigma_e(n)$, the sum of even divisors of n, is given by*

$$\frac{\varsigma(s)\varsigma(s-1)}{2^{s-1}} = \sum_{n=1}^{\infty} \frac{\sigma_e(n)}{n^s}$$

Proof. Since $\sigma_e(n) = \sigma(n) - \sigma_o(n)$, we simply subtract the generating function for $\sigma_o(n)$ from the generating function for $\sigma(n)$:

$$\varsigma(s)\varsigma(s-1) - \varsigma(s)\varsigma(s-1)(1 - \frac{1}{2^{s-1}}) = \frac{\varsigma(s)\varsigma(s-1)}{2^{s-1}} \quad \square$$

Exponential Relations for Partition Functions

In this section, we will derive some interesting exponential relations for several partition functions. Whether these results prove to be of any value is something that other mathematicians will have to decide. I present them here because they are sufficiently interesting to preserve for future investigation.

Let us define a "weighting function" $\omega_k(n) = \begin{cases} 1, & \text{if } k \text{ divides } n \\ 0, & \text{if } k \text{ does not divide } n \end{cases}$, for integer $n \geq 1$. Then

$\frac{\sigma(n)}{n} = \frac{\sum_{k=1}^{n} \omega_k(n) \cdot k}{n}$. This is clearly a kind of weighted average of the numbers 1, 2, ..., n, where the weights are 1 or 0 according to whether $k \in [n]$ does or does not divide n, respectively. We can define generating functions for these "weighted averages" that have a nice relationship to the partition functions. Let us define the following ordinary power series generating functions:

(1) $P(x) = \sum_{n=0}^{\infty} \pi(n) x^n$, where $\pi(n)$ is the number of partitions of n.

(2) $P_d(x) = \sum_{n=0}^{\infty} \pi_d(n) x^n$, where $\pi_d(n)$ is the number of partitions of n into distinct parts.

(3) $P_o(x) = \sum_{n=0}^{\infty} \pi_o(n) x^n$, where $\pi_o(n)$ is the number of partitions of n into odd parts.

(4) $P_e(x) = \sum_{n=0}^{\infty} \pi_e(n) x^n$, where $\pi_e(n)$ is the number of partitions of n into even parts.

(5) $D(x) = \sum_{n=1}^{\infty} \frac{\sigma(n)}{n} x^n$, where $\sigma(n)$ is the sum of divisors of n.

(6) $D_o(x) = \sum_{n=1}^{\infty} \frac{\sigma_o(n)}{n} x^n$, where $\sigma_o(n)$ is the sum of odd divisors of n.

(7) $D_e(x) = \sum_{n=1}^{\infty} \frac{\sigma_e(n)}{n} x^n$, where $\sigma_e(n)$ is the sum of even divisors of n.

These generating functions have remarkably simple and beautiful relationships:

Theorem 5.16. $D(x) = D_o(x) + D_e(x)$, and $P(x) = e^{D(x)}$, $P_d(x) = e^{D_o(x)}$, $P_e(x) = e^{D_e(x)}$.

Note that an immediate consequence of Theorem 5.16 is that
$$P(x) = e^{D(x)} = e^{D_o(x) + D_e(x)} = e^{D_o(x)} \cdot e^{D_e(x)} = P_d(x) \cdot P_e(x).$$

Also, by symmetry we suspect that $P_d(x) = P_o(x)$. (This is true, and it is known as Euler's theorem for partitions, although Euler stated it somewhat differently than we did here.) See Theorem 5.17.

Proof. First, we show that $D(x) = D_o(x) + D_e(x)$.

$$D(x) = \sum_{n=1}^{\infty} \frac{\sigma(n)}{n} x^n$$

$$= \sum_{n=1}^{\infty} \frac{\sigma_o(n) + \sigma_e(n)}{n} x^n$$

$$= \sum_{n=1}^{\infty} (\frac{\sigma_o(n)}{n} + \frac{\sigma_e(n)}{n}) x^n$$

$$= \sum_{n=1}^{\infty} \frac{\sigma_o(n)}{n} x^n + \sum_{n=1}^{\infty} \frac{\sigma_e(n)}{n} x^n$$

$$= D_o(x) + D_e(x) \quad \square$$

Next, by Theorem 5.11, we have

$$n \cdot \pi_d(n) = \sum_{j=0}^{n-1} \pi_d(j) \sigma_o(n-j)$$

$$= \sum_j \pi_d(j) \sigma_o(n-j) \text{ since } \sigma_o(0) = 0.$$

By the Cauchy product rule, the generating function for $\sum_j \pi_d(j) \sigma_o(n-j)$ is $P_d(x) \cdot x \cdot \frac{dD_o(x)}{dx}$. The generating function for $n \cdot \pi_d(n)$ is $x \cdot \frac{dP_d(x)}{dx}$. Therefore, $x \cdot \frac{dP_d(x)}{dx} = x \cdot P_d(x) \cdot \frac{dD_o(x)}{dx}$, or simply $\frac{dP_d(x)}{dx} = P_d(x) \cdot \frac{dD_o(x)}{dx}$. This differential equation is of the type *variables separable*. It has the general solution, with boundary conditions $P_d(0) = 1$ and $D_o(0) = 0$, of $P_d(x) = e^{D_o(x)}$. The analogous result, $P(x) = e^{D(x)}$, can be proved analogously to the foregoing proof. Next, since $P(x) = e^{D_e(x)} \cdot P_d(x)$ by using the result just proved, we have

$$e^{D_e(x)} = \frac{P(x)}{P_d(x)} = \frac{(1+x+x^2+...)}{(1+x)} \frac{(1+x^2+x^4+...)}{(1+x^2)} \cdots$$

$$= (1+x^2+x^4+x^6+...)(1+x^4+x^8+x^{12}+...)...$$

$$= P_e(x).$$

The penultimate result is obtained by direct polynomial division. $\square$

One potential benefit in using the exponential relationships for the partition generating functions is that they are very easy to work with. We can easily see, for example, that the following relations are true:

$$P(x) = e^{D(x)}$$
$$P^k(x) = e^{k \cdot D(x)}$$
$$P^{-1}(x) = e^{-D(x)} \Rightarrow P(x)P^{-1}(x) = 1$$
$$D(x) = \log_e(P(x))$$

In terms of generating functions, each of these relations can be represented as a formal power series (e.g., using the Taylor series expansion of e^x). Then, if we perform a huge amount of algebra, we will see that both sides are equal. We can now easily prove Euler's theorem:

Theorem 5.17 (Euler). *The number of partitions of n into distinct parts is the same as the number of partitions of n into odd parts.*

Proof.

$$P_d(x) = \frac{P(x)}{P_e(x)}$$

$$= \frac{\left(\frac{1}{1-x}\right)\left(\frac{1}{1-x^2}\right)\left(\frac{1}{1-x^3}\right)\left(\frac{1}{1-x^4}\right)\cdots}{\left(\frac{1}{1-x^2}\right)\left(\frac{1}{1-x^4}\right)\left(\frac{1}{1-x^6}\right)\left(\frac{1}{1-x^8}\right)\cdots}$$

$$= \left(\frac{1}{1-x}\right)\left(\frac{1}{1-x^3}\right)\left(\frac{1}{1-x^5}\right)\cdots \text{ (by cancellation)}$$

$$= P_o(x) \square$$

This proof is very similar to Euler's original proof, but we managed to circumvent the necessary stroke of genius in Euler's proof by "climbing the mountain from the other side" ([8], pp. 164-165).

Here is one final theorem, almost certainly known, which we can prove by starting with an exponential representation:

Theorem 5.18. $n \cdot \pi_e(n) = \sum_{k=0}^{\lfloor (n-1)/2 \rfloor} \pi_e(2k)\sigma_e(n-2k)$, where $\pi_e(n)$ is the number of partitions of n into even parts, and $\sigma_e(n)$ is the sum of even divisors of n. We define $\pi_e(0) \equiv 1$.

Proof.

$$P_e(x) = e^{D_e(x)}$$

$$x\frac{dP_e(x)}{dx} = xe^{D_e(x)} \cdot \frac{dD_e(x)}{dx}, \text{ by formal differentiation}$$

$$\sum_{n=0}^{\infty} n\pi_e(n)x^n = xP_e(x)\frac{dD_e(x)}{dx}, \text{ by definitions}$$

$$= \sum_{n=0}^{\infty} \pi_e(n)x^n \cdot \sum_{n=1}^{\infty} \sigma_e(n)x^n, \text{ by substitution}$$

$$= \sum_{n=0}^{\infty} \pi_e(n)x^n \cdot \sum_{n=0}^{\infty} \sigma_e(n)x^n, \text{ since } \sigma_e(0) = 0$$

$$\sum_{n=0}^{\infty} (\sum_{j=0}^{n} \pi_e(j)\sigma_e(n-j))x^n, \text{ by the Cauchy product rule}$$

Equating coefficients of x^n on both sides of the equation gives

$$n \cdot \pi_e(n) = \sum_{j=0}^{n} \pi_e(j)\sigma_e(n-j)$$

$$= \sum_{j=0}^{n-1} \pi_e(j)\sigma_e(n-j), \text{ since } \sigma_e(0) = 0$$

$$= \sum_{k=0}^{\lfloor (n-1)/2 \rfloor} \pi_e(2k)\sigma_e(n-2k), \text{ since } \pi_e(n) = 0 \text{ for } n \text{ odd.} \quad \square$$

Multiplicative Inverse of the Distinct Occupancy Generating Function

The distinct occupancy generating function, $\Delta(x, t)$, enumerates the number of ways to distribute objects of arbitrary specification into identical boxes, with no box empty, such that the occupancies of the boxes are distinct. In other words, no two boxes are identical with regard to the objects contained in each box. We will show that the multiplicative inverse of the distinct occupancy generating function is $G(x, -t)$, where $G(x, t)$ is the generating function for general, or unrestricted, box occupancies:

$$\Delta(x, t) \cdot G(x, -t) = 1$$

Theorem 5.19. *The multiplicative inverse of the Distinct Occupancy Generating Function (Theorem 5.1) is $\Delta^{-1}(x, t) = G(x, -t)$, where $G(x, t)$ is given by Theorem 2.8.*

Proof. The proof is an easy observation that $\Delta(x, t) = \dfrac{1}{G(x, -t)}$ by substitution of $-t$ for t in Theorem 2.8 and comparison with Theorem 5.1. Hence, we have $\Delta(x, t) \cdot G(x, -t) = 1$. □

Theorem 5.20. *For positive integer n,* $\delta_n(t) * g_n(-t) = \begin{cases} 1, & n = 1 \\ 0, & \text{otherwise} \end{cases}$. *Furthermore, we have, for positive integer $n > 1$, $\sum_{d \mid n} \delta(d) \cdot g_{\frac{n}{d}}(-1) = 0$.*

Proof. By Theorem 5.19, $\Delta(x, t) \cdot G(x, -t) = 1$. Multiplying the two generating functions gives

$$\left(\sum_{n=1}^{\infty} \delta_n(t) x^{\log n} \right) \left(\sum_{n=1}^{\infty} g_n(-t) x^{\log n} \right) = 1,$$

which implies $\delta_n(t) * g_n(-t) = \begin{cases} 1, & n = 1; \\ 0, & \text{otherwise}. \end{cases}$

Hence, by the convolution product in the last result with $t = 1$, we have $\delta_n(t) * g_n(-t) = \sum_{d \mid n} \delta(d) \cdot g_{\frac{n}{d}}(-1) = 0$, whenever $n > 1$. We also have, of course, the trivial result that $\delta_1(t) * g_1(-t) = 1$, but this case is not very interesting. □

Example 5.21. Let $n = 12 = 2^2 \cdot 3$. Here, $n > 1$. So, using tables of distribution polynomials (Appendix A and Table 5.1, where $\delta(n) = \delta_n(1)$), we have

$$\sum_{d|12} \delta(d) \cdot g_{\frac{12}{d}}(-1) = \delta(1)g_{12}(-1) + \delta(2)g_6(-1) + \delta(3)g_4(-1) + \delta(4)g_3(-1) + \delta(6)g_2(-1) + \delta(12)g_1(-1)$$

$$= (1)(0) + (1)(0) + (1)(0) + (1)(-1) + (2)(-1) + (3)(1)$$
$$= 0 + 0 + 0 - 1 - 2 + 3$$
$$= 0.$$

Corollary 5.22. *For positive integer* $n > 1$, $\delta(n) = -\sum_{\substack{d|n \\ d \neq n}} \delta(d) \cdot g_{\frac{n}{d}}(-1)$.

Proof. Simply solve for $\delta(n)$ in Theorem 5.20.

Since we can use Corollary 5.22 recursively for each $\delta(d)$ by back-substitution, Corollary 5.22 effectively tells us that we can represent the number of distinct occupancy distributions $\delta(n)$, for a given specification number $n > 1$, in terms of general occupancy polynomials $g_k(t)$ evaluated at $t = -1$, or $g_k(-1)$.

Theorem 5.23. *Let $\pi_d(n)$ be the number of partitions of a non-negative integer n into distinct parts. Let $\pi(n, t)$ be the partition polynomial for the non-negative integer n. The coefficient of t^k in $\pi(n,t)$ is the number of partitions of n into k parts. Also, define $\pi_d(0) = 1$ and $\pi(0, -1) = 1$. (The reason we allow n = 0 is to simplify the notation.) Then we have the following identity for integer partitions:*

$$\sum_{r=0}^{k} \pi_d(r) \cdot \pi(k-r, -1) = 0.$$

Proof. For positive integer $n > 1$, $\sum_{d|n} \delta(d) \cdot g_{\frac{n}{d}}(-1) = 0$ by Theorem 5.20. Let $n = p^k$ for some prime number p. Then, any divisor d of n will have the form $d = p^r$. So, $\delta(d) = \delta(p^r)$, which is the number of distinct partitions of r, or $\pi_d(r)$. Similarly, $g_{\frac{n}{d}}(-1) = g_{p^{k-r}}(-1) = \pi(k-r, -1)$. □

Example 5.24. Let $k = 5$. Let us demonstrate that $\sum_{r=0}^{5} \pi_d(r) \cdot \pi(5-r, -1) = 0$. First, let us recall the first five partition polynomials:

$$\pi(1, t) = t$$
$$\pi(2, t) = t + t^2$$
$$\pi(3, t) = t + t^2 + t^3$$
$$\pi(4, t) = t + 2t^2 + t^3 + t^4$$
$$\pi(5, t) = t + 2t^2 + 2t^3 + t^4 + t^5$$

For example, the partition polynomial $\pi(4, t) = t + 2t^2 + t^3 + t^4$ tells us that there is one partition of 4 into one part, two partitions of 4 into two parts, one partition of 4 into three parts, and one partition of 4 into four parts. The coefficient of t^k gives the number of partitions of 4 into k parts. Now let us verify that $\sum_{r=0}^{5} \pi_d(r) \cdot \pi(5-r, -1) = 0$. We have

$$\pi_d(0)\pi(5, -1) + \pi_d(1)\pi(4, -1) + \pi_d(2)\pi(3, -1) + \pi_d(3)\pi(2, -1) + \pi_d(4)\pi(1, -1) + \pi_d(5)\pi(0, -1)$$
$$= (1)(-1) + (1)(1) + (1)(-1) + (2)(0) + (2)(-1) + (3)(1)$$
$$= -1 + 1 - 1 + 0 - 2 + 3$$
$$= 0$$

Therefore, $\sum_{r=0}^{5} \pi_d(r) \cdot \pi(5-r, -1) = 0$ as expected. □

A Generalization of Euler's Partition Theorem for Vector Partitions

Theorem 5.17, due to Leonhard Euler, tells us that the number of positive integer partitions of a positive integer n into odd parts is equinumerous with the number of partitions of n into distinct parts. We can write this statement more compactly as $p(n\,|\,\text{each part distinct}) = p(n\,|\,\text{each part odd})$. Euler's theorem can be generalized to vector partitions. Vector partitions were briefly discussed in Chapter 2, in the section entitled "Equivalent Problems."

In Chapter 2, we noted that vector partitions are isomorphic to distributions of objects, such as colored balls, into identical boxes with no box empty. We also noted that these problems are isomorphic to unordered factorizations of a positive integer into non-unit, positive factors. To prove a generalization of Euler's theorem for vector partitions, we will adopt the equivalent viewpoint of distributing objects into identical boxes. This approach will allow us to prove the theorem using the machinery of the logarithmic generating function.

The number of distributions of objects of specification n into identical boxes, such that each box contains a *distinct* set of objects, such as colored balls, is given by Theorem 5.1:

$$\Delta(x,t) = \sum_{n=1}^{\infty} \delta_n(t) x^{\log n} = \prod_{d>1}(1+tx^{\log d}).$$

Note that if $n = p_1^{r_1} p_2^{r_2} \cdots p_k^{r_k}$, then $\delta(n) = \delta_n(1)$ counts the total number of distributions of r_1 balls of one color, r_2 balls of a second color, ..., and r_k balls of a kth color into identical boxes, with no box empty, such that the occupancies of the boxes are distinct. This is isomorphic to the statement that we are partitioning the vector $(r_1, r_2, ..., r_k)$ into a sum of *distinct* non-zero vector parts.

Now suppose that we want to distribute objects (colored balls) of specification n into identical boxes, with no box empty, such that each box contains an odd number of balls of *at least* one color. We want to build a generating function to count the number, $O(n)$, of such distributions. To do this, we note that if a box contains an even number of balls for each color that is in the box, then the specification number for that box is a perfect square. For example, if a box contains the multiset of colored balls $S = \{r, r, w, w, w, w, b, b\} = \{2, 2, 3, 3, 3, 3, 5, 5\}$, then the specification number is $n = 2^2 \cdot 3^4 \cdot 5^2 = (2 \cdot 3^2 \cdot 5)^2 = m^2$. As a vector partition, this example would be written as $v = (r, w, b) = (2, 4, 2)$, since we have 2 red balls, 4 white balls, and 2 blue balls. So, to build a generating function for $O(n)$, we must exclude the possibility of a box having an even number of balls for each of the colors that are found within that box. Equivalently, if we think in terms of unordered factorizations of a positive integer into non-unit, positive factors, we want to exclude those factors that are perfect squares. Thus, a logarithmic generating function for enumerating $O(n)$ is given by

$$O(x,t) = \sum_{n=1}^{\infty} O_n(t) x^{\log n}$$
$$= (1+tx^{\log 2} + t^2 x^{\log(2\cdot 2)} + t^3 x^{\log(2\cdot 2\cdot 2)} + ...)(1+tx^{\log 3} + t^2 x^{\log(3\cdot 3)} + t^3 x^{\log(3\cdot 3\cdot 3)} + ...)$$
$$\cdot (1+tx^{\log 5} + t^2 x^{\log(5\cdot 5)} + t^3 x^{\log(5\cdot 5\cdot 5)} + ...)(1+tx^{\log 6} + t^2 x^{\log(6\cdot 6)} + t^3 x^{\log(6\cdot 6\cdot 6)} + ...)\cdots$$
$$= \frac{1}{(1-tx^{\log 2})(1-tx^{\log 3})(1-tx^{\log 5})(1-tx^{\log 6})(1-tx^{\log 7})(1-tx^{\log 8})(1-tx^{\log 10})\cdots}.$$

Note that we have excluded the terms with "square factors," such as $\frac{1}{(1-tx^{\log 4})}$, $\frac{1}{(1-tx^{\log 9})}$, ..., $\frac{1}{(1-tx^{\log k^2})}$, ... We are now ready to prove our generalization of Euler's partition theorem for vector partitions.

Theorem 5.25 (Generalization of Euler's partition theorem for vector partitions). *The number of vector partitions of a non-zero k-tuple of nonnegative integers into non-zero distinct parts is equal to the number of vector partitions of the k-tuple into non-zero parts such that each part has at least one odd component. (Note that a "zero part" is a vector of zeros (0, 0, ... , 0).)*

Proof. The proof is very similar to Euler's proof, but we use the logarithmic generating functions that we presented and constructed in the previous sections. By Theorem 5.1, we have

$$\Delta(x,t) = \prod_{d>1}(1+tx^{\log d})$$

$$(1+tx^{\log 2})(1+tx^{\log 3})(1+tx^{\log 4})\cdots$$

$$= \frac{(1+tx^{\log 2})(1-tx^{\log 2})(1+tx^{\log 3})(1-tx^{\log 3})(1+tx^{\log 4})(1-tx^{\log 4})\cdots}{(1-tx^{\log 2})(1-tx^{\log 3})(1-tx^{\log 4})\cdots}$$

$$= \frac{(1-t^2 x^{\log(2\cdot 2)})(1-t^2 x^{\log(3\cdot 3)})(1-t^2 x^{\log(4\cdot 4)})\cdots}{(1-tx^{\log 2})(1-tx^{\log 3})(1-tx^{\log 4})\cdots}.$$

At this point, it may appear as though we are "stuck," because we have terms with t^2 in the numerator and t in the denominator, so nothing cancels. However, if we let $t=1$, then we get

$$\Delta(x,1) = \frac{(1-x^{\log(2\cdot 2)})(1-x^{\log(3\cdot 3)})(1-x^{\log(4\cdot 4)})\cdots}{(1-x^{\log 2})(1-x^{\log 3})(1-x^{\log 4})\cdots}$$

$$= \frac{(1-x^{\log 4})(1-x^{\log 9})(1-x^{\log 16})\cdots}{(1-x^{\log 2})(1-x^{\log 3})(1-x^{\log 4})\cdots}$$

$$= \frac{1}{(1-x^{\log 2})(1-x^{\log 3})(1-x^{\log 5})\cdots}$$

$$= O(x,1).$$

Since $\Delta(x,1) = O(x,1)$, we have $\sum_{n=1}^{\infty} \delta_n(1) x^{\log n} = \sum_{n=1}^{\infty} O_n(1) x^{\log n}$, or $\sum_{n=1}^{\infty} \delta(n) x^{\log n} = \sum_{n=1}^{\infty} O(n) x^{\log n}$.

Equating coefficients of $x^{\log n}$ on both sides of the equation gives the desired result: $\delta(n) = O(n)$. □

Building General Occupancy Polynomials from Distinct Occupancy Polynomials

While it may not be surprising, the object distribution polynomials for distinct occupancy, $\delta_n(t)$, can be used as building blocks to construct object distribution polynomials for general, or arbitrary, occupancy, $g_n(t)$. The challenge, of course, is to show how this can be done. The following theorem gives us one way to represent $g_n(t)$ in terms of distinct occupancy polynomials.

Theorem 5.26. *For positive integer $n > 1$, $g_n(t) = \sum_{n = f_1 f_2 \cdots f_k} (-1)^k \delta_{f_1}(-t) \delta_{f_2}(-t) \cdots \delta_{f_k}(-t)$, where the sum is taken over all ordered factorizations of n, excluding 1 as a factor.*

Proof. By Theorem 5.20, we have $\delta_n(t) * g_n(-t) = \begin{cases} 1, & n = 1 \\ 0, & \text{otherwise} \end{cases}$. So, for positive integer $n > 1$, $\delta_n(t) * g_n(-t) = 0$. By definition of the convolution product, and interchanging the roles of t and $-t$, this means that $\sum_{d \mid n} \delta_d(-t) g_{\frac{n}{d}}(t) = 0$. Solving for $g_n(t)$, and using the fact that $\delta_1(t) = 1$, we get $g_n(t) = -\sum_{\substack{d \mid n \\ d \neq 1}} \delta_d(-t) g_{\frac{n}{d}}(t)$. Now, we can use the equation $g_n(t) = -\sum_{\substack{d \mid n \\ d \neq 1}} \delta_d(-t) g_{\frac{n}{d}}(t)$ and back-substitute for $g_{\frac{n}{d}}(t)$ into the same equation, and continue with repeated back-substitution until we finally obtain the desired result: $g_n(t) = \sum_{n = f_1 f_2 \cdots f_k} (-1)^k \delta_{f_1}(-t) \delta_{f_2}(-t) \cdots \delta_{f_k}(-t)$. □

Example 5.27. The ordered factorizations of $n = 12$, excluding unit factors, are 12, $2 \cdot 6$, $6 \cdot 2$, $3 \cdot 4$, $4 \cdot 3$, $2 \cdot 2 \cdot 3$, $2 \cdot 3 \cdot 2$, and $3 \cdot 2 \cdot 2$. (There are 8 ordered factorizations of 12 using non-unit factors.) So, by Theorem 5.26, we have

$$\begin{aligned} g_{12}(t) &= (-1)^1 \delta_{12}(-t) + (-1)^2 \delta_2(-t) \delta_6(-t) + (-1)^2 \delta_6(-t) \delta_2(-t) \\ &+ (-1)^2 \delta_3(-t) \delta_4(-t) + (-1)^2 \delta_4(-t) \delta_3(-t) + (-1)^3 \delta_2(-t) \delta_2(-t) \delta_3(-t) \\ &+ (-1)^3 \delta_2(-t) \delta_3(-t) \delta_2(-t) + (-1)^3 \delta_3(-t) \delta_2(-t) \delta_2(-t). \end{aligned}$$

This simplifies to $g_{12}(t) = -\delta_{12}(-t) + 2\delta_2(-t)\delta_6(-t) + 2\delta_3(-t)\delta_4(-t) - 3\delta_2(-t)\delta_2(-t)\delta_3(-t)$. Knowing that (Table 5.1) $\delta_2(t) = \delta_3(t) = \delta_4(t) = t$, $\delta_6(t) = t + t^2$, and $\delta_{12}(t) = t + 2t^2$, we can easily verify that $g_{12}(t) = t + 2t^2 + t^3$, which is correct.

We can also represent the distinct occupancy polynomials in terms of the general occupancy polynomials.

Theorem 5.28. *For positive integer* $n > 1$, $\delta_n(t) = \sum_{n=f_1 f_2 \cdots f_k} (-1)^k g_{f_1}(-t) g_{f_2}(-t) \cdots g_{f_k}(-t)$, *where the sum is taken over all ordered factorizations of n, excluding 1 as a factor.*

Proof. We proved Theorem 5.26 using the *identity* $\delta_n(t) * g_n(-t) = 0$ for positive integer $n > 1$. Since the convolution product is commutative, and since we can replace the variable t by $-t$, we have the equivalent relation $g_n(t) * \delta_n(-t) = 0$. Thus, we can interchange the roles of g and δ in the proof of Theorem 5.26 to immediately obtain the result stated as Theorem 5.28. □

Comment. Note the beautiful symmetry in Theorems 5.26 and 5.28. The representation of $g_n(t)$ in terms of the δ's is identical in form to the representation of $\delta_n(t)$ in terms of the g's. The representations are invariant under the interchange of g and δ.

As an aid to determining the ordered factorizations of n, which we need for Theorems 5.26 and 5.28, it is useful to know how many ordered factorizations of n exist. To this end, the following recurrence relation allows us to count the number of *ordered* factorizations of a positive integer n, excluding unit factors.

Theorem 5.29. *For positive integer n, the number of* ordered *factorizations of n, excluding unit factors, denoted by* $F(n)$, *is given by the recurrence relation* $F(n) = \sum_{\substack{d|n \\ d \neq n}} F(d)$, *where* $F(1) = 1$. (For

$n > 1$, we exclude unit factors. For $n = 1$, we need to define $F(1) = 1$ so that the recurrence relation works.)

Proof. In any ordered factorization of $n > 1$, excluding unit factors, the first factor can be any number d such that d divides n. By appending to this d to all possible ordered factorizations of n/d, we obtain the recurrence formula for $F(n)$: $F(n) = \sum_{\substack{d|n \\ d \neq n}} F(d)$, with $F(1) = 1$. □

Example 5.30. Find the number of ordered factorizations of 12. We know that there are, in fact, eight *ordered* factorizations of 12, namely 12, $2 \cdot 6$, $6 \cdot 2$, $3 \cdot 4$, $4 \cdot 3$, $2 \cdot 2 \cdot 3$, $2 \cdot 3 \cdot 2$, and $3 \cdot 2 \cdot 2$. Using the recurrence relation given in Theorem 5.29, we have

$$F(12) = \sum_{\substack{d|12 \\ d \neq 12}} F(d)$$
$$= F(1) + F(2) + F(3) + F(4) + F(6)$$
$$= 1 + 1 + 1 + 2 + 3$$
$$= 8.$$

Exercises

1. Use Theorem 5.11 to show that $\pi_d(5) = 3$. What are the partitions of 5 into distinct parts?

2. In how many different ways can you distribute three red balls and three blue balls into precisely three identical boxes with no box empty such that no two boxes contain the same set of colored balls?

3. Calculate a Maclaurin series expansion of $e^{D(x)}$ for several terms and show that the result agrees with Theorem 5.16: $P(x) = e^{D(x)}$.

4. Use the polynomials in Appendix A and Table 5.1 to show that $\delta_{36}(t) * g_{36}(-t) = 0$.

5. Use Theorem 5.29 to calculate the number of ordered factorizations of 18 in non-unit factors.

Chapter 6. Distributions in Identical Boxes with Even Color Occupancies

In how many different ways can we distribute objects of arbitrary specification, such as colored balls, into *identical* boxes such that the occupancies of the boxes have *even* color occupancy? A simple example will illustrate what we mean. Suppose that we have six red balls, eight white balls, and two blue balls. The balls of a given color, such as red, are identical among themselves. In how many ways can we distribute these colored balls into k identical boxes, with no box empty, such that each color that occurs within a box occurs with even color replication? Figure 6.1 shows one such distribution for $k = 4$. It is not necessary for a box to contain each possible color. A box can contain balls of any arbitrary colors. However, whatever colors do occur within a box must occur with *even* replication. Hence, each box can contain either 0, 2, 4, 6, ... *red* balls, 0, 2, 4, 6, ... *white* balls, 0, 2, 4, 6, ... *blue* balls, and so on for all possible colors.

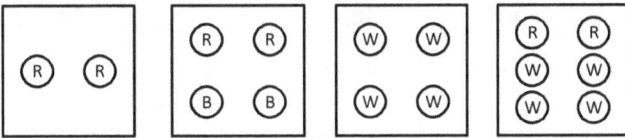

Figure 6.1. A distribution of colored balls in four boxes with even color replication in each box.

The distribution shown in Figure 6.1 can also be interpreted as a *vector partition* of the form $(r, w, b) = (6, 8, 2) = (2, 0, 0) + (2, 0, 2) + (0, 4, 0) + (2, 4, 0)$.

In the following discussion, we restrict the distribution problem to the requirement that no box can be empty. (The case where empty boxes are allowed is a trivial consequence of our present problem, obtained by simple summation over the number of boxes.) Note that the order of the balls within a box is immaterial. Within a given box, we can rearrange the order of the balls without changing its occupancy. Also, since the boxes themselves are identical, the order of the boxes is immaterial, because we cannot tell the empty boxes apart. Only when we place objects into the boxes do they acquire an "identity."

A Generating Function

To count, or enumerate, the number distributions of colored balls into identical boxes such that each box has even color occupancy, we first build an appropriate generating function. Theorem 2.8 provides a generating function, called the *logarithmic generating function*, for solving the general, or unrestricted occupancy, distribution problem:

$$G(x,t) = \prod_{d>1} \frac{1}{(1-tx^{\log d})} = \sum_{n=1}^{\infty} g_n(t) x^{\log n}. \qquad (6.1)$$

For our present purpose, we would like to modify the logarithmic generating function (6.1) to allow us to enumerate the distributions with even color replication in each box. In (6.1), the coefficient of $x^{\log n}$ is a polynomial $g_n(t)$. The polynomial $g_n(t)$ is called the *object distribution polynomial* for a set of colored balls of *specification n* (Definition 2.1). This polynomial counts the number of ways to distribute colored balls of specification n into identical boxes with no box empty. In particular, when we expand the generating function (6.1), and collect like terms, we find that each term of the form $t^k x^{\log n^k} = t^k x^{k \log n}$ represents colored balls of specification n replicated k times, with one complete replication in each of the k boxes. So, we must somehow modify (6.1) to obtain a generating function that only counts even color replications. Before we derive such a generating function, let us digress momentarily to remember the following definition:

Definition 6.1. A multiset of colored balls, $S = \{p_1^{n_1}, p_2^{n_2}, ..., p_k^{n_k}\}$, where each color is represented by a distinct prime p_j, and each color occurs with repetition n_j, is encoded by a *specification number n*, where $n = p_1^{n_1} p_2^{n_2} ... p_k^{n_k}$.

The specification number is not unique, because different primes can be chosen to represent the different colors. However, for any choice of distinct primes p_j, any two specifications $n = p_1^{n_1} p_2^{n_2} \ldots p_k^{n_k}$, for fixed $n_1, n_2, \ldots, n_k$ and k, are said to have the same *form*. This means that specifications of the same form will have identical distributions, since we are effectively doing nothing more than relabeling the balls.

Example 6.2. Two red balls, one white ball, and three blue balls can be represented by the multiset $S = \{red, red, white, blue, blue, blue\}$. If we let the prime number 2 represent the color *red*, let the prime 3 represent the color *white*, and let the prime 5 represent the color *blue*, then we can write the set S as $S = \{2^2, 3^1, 5^3\}$. Then the *specification n* for this set of colored balls is $n = 2^2 \cdot 3 \cdot 5^3 = 1500$. By the Fundamental Theorem of Arithmetic, we can recover the original primes by prime factorizing the specification number.

The following lemma will be used in the derivation of Theorem 6.4.

Lemma 6.3. *Let $\tau(n)$ denote the total number of positive integer divisors of a positive integer n. Then $\tau(n)$ is odd if and only if n is a perfect square.*

Proof. If n is a square, then $n = p_1^{\alpha_1} p_2^{\alpha_2} \cdots p_k^{\alpha_k} = p_1^{2\beta_1} p_2^{2\beta_2} \cdots p_k^{2\beta_k} = (p_1^{\beta_1} p_2^{\beta_2} \cdots p_k^{\beta_k})^2$, and $\tau(n) = (2\beta_1 + 1)(2\beta_2 + 1) \cdots (2\beta_k + 1) = odd$. Conversely, if $\tau(n)$ is odd, then $\tau(n)$ has the form $\tau(n) = odd \cdot odd \cdots odd = (2\beta_1 + 1)(2\beta_2 + 1) \cdots (2\beta_k + 1)$, which implies that $n = p_1^{2\beta_1} p_2^{2\beta_2} \cdots p_k^{2\beta_k} = (p_1^{\beta_1} p_2^{\beta_2} \cdots p_k^{\beta_k})^2 = m^2$. □

Theorem 6.4. Let $\varepsilon_n(t)$ be the distribution polynomial for distributing colored balls of specification n into identical boxes, with no boxes empty, such that each color that occurs within a given box occurs with even replication. The logarithmic generating function for $\varepsilon_n(t)$ is given by

$$E(x,t) = \prod_{d>1} \frac{1}{(1 - tx^{2\log d})} = \sum_{n=1}^{\infty} \varepsilon_n(t) x^{\log n}. \qquad (6.2)$$

Proof. We want to build a generating function to calculate the number of distributions for this problem. In the generating function, each term of the form $t^k x^{\log n^k} = t^k x^{k \log n}$ represents colored balls (objects) of specification n replicated k times—one replication in each of the k identical boxes. For example, $t^2 x^{\log 2^2} = t^2 x^{2 \log 2} = (tx^{\log 2})(tx^{\log 2})$ represents a colored ball of specification $n = 2$, say one *red* ball, replicated once in each of two boxes: {2} and {2}, or {red} and {red}. Remember, the exponent of t is the number of boxes. A term like $t^1 x^{\log 4^1} = t^1 x^{1 \log 4} = tx^{\log 4}$ represents colored balls of specification $n = 4 = 2^2$, say two red balls, replicated once in one box: {2, 2} or {red, red}. Consider a general term like $t^k x^{\log n^k}$. Suppose that the specification number n has the form $n = p_1^{\alpha_1} p_2^{\alpha_2} \cdots p_k^{\alpha_k}$, for distinct primes $p_1, p_2, ..., p_k$. Here, we interpret n as representing a set of α_1 balls of "color" p_1, α_2 balls of color p_2, ..., and α_k balls of color p_k. Since the exponent of t is k, we have k boxes. Each one of the k boxes will contain a complete set of colored balls of specification n. So each box contains a total of $\alpha_1 + \alpha_2 + ... + \alpha_k$ balls. In order for a box to contain colored balls with *even* color replication within that box, we must have $\alpha_j = even$ for each $j = 1, 2, ..., k$. Now consider the number-theoretic function $\tau(n)$, where $\tau(n)$ is the number of positive integral divisors of n. If $n = p_1^{\alpha_1} p_2^{\alpha_2} \cdots p_k^{\alpha_k}$, for distinct primes $p_1, p_2, ..., p_k$, then $\tau(n) = (\alpha_1 + 1)(\alpha_2 + 1) \cdots (\alpha_k + 1)$. If $\tau(n)$ is odd, then $\alpha_j = even$ for all j. Therefore, the logarithmic generating function that solves our distribution problem of even color occupancy in each box is given by

$$\prod_{\substack{d>1 \\ \tau(d) odd}} \frac{1}{(1-tx^{\log d})} = \sum_{n=1}^{\infty} \varepsilon_n(t) x^{\log n} \qquad (6.3)$$

By Lemma 6.3, $\tau(n)$ is odd if and only if n is a perfect square. Therefore, (6.3) becomes

$$\sum_{n=1}^{\infty} \varepsilon_n(t) x^{\log n} = \prod_{\substack{d>1 \\ \tau(d) odd}} \frac{1}{(1-tx^{\log d})}$$

$$= \prod_{\substack{d>1 \\ d=m^2}} \frac{1}{(1-tx^{\log d})}$$

$$= \prod_{m>1} \frac{1}{(1-tx^{\log m^2})}$$

$$= \prod_{m>1} \frac{1}{(1-tx^{2\log m})} \qquad \square$$

Remark 6.5. Note that the only way that we can distribute colored balls into boxes such that each box has even color replication is if we have an even number of balls of each color. This means that α_j must be even for each j, or, equivalently, that $n = p_1^{\alpha_1} p_2^{\alpha_2} \cdots p_k^{\alpha_k} = p_1^{2\beta_1} p_2^{2\beta_2} \cdots p_k^{2\beta_k} = (p_1^{\beta_1} p_2^{\beta_2} \cdots p_k^{\beta_k})^2$. Thus, the specification number n must be a perfect square. If n is not a perfect square, then $\varepsilon_n(t) = 0$. This observation can be used to simplify many computations. We state this useful observation formally as Lemma 6.6:

Lemma 6.6. *If n is not a perfect square, then $\varepsilon_n(t) = 0$.*

A Recurrence Relation

In a certain sense, we could argue that the distribution problem for distinct occupancy is completely solved. The entire solution is contained in Theorem 6.4. Unfortunately, expanding the generating function, by multiplying its terms, involves a great many multiplications. What we really need, for the sake of computational efficiency, is a *recurrence relation*. The recurrence relation will allow us to calculate distribution polynomials $\varepsilon_n(t)$ in terms of other distribution polynomials for smaller values of n. The following theorem is central to this effort:

Theorem 6.7. *A recurrence relation for $\varepsilon_n(t)$ is given by*

$$\varepsilon_n(t) \cdot \log n = 2 \cdot \sum_{d>1} \sum_{j \geq 1} \varepsilon_{\frac{n}{d^{2j}}}(t) \cdot t^j \cdot \log d. \tag{6.4}$$

Here, we sum over all square divisor d^2 of n, $d \neq 1$, taken to the jth power for $j \geq 1$. We define $\varepsilon_1(t) = 1$.

Proof. By Theorem 6.4, we have $E(x,t) = \prod_{d>1} \dfrac{1}{(1-tx^{2\log d})} = \sum_{n=1}^{\infty} \varepsilon_n(t) x^{\log n}$. Taking the natural logarithm of both sides of the equation gives

$$\log(\sum_{n=1}^{\infty} \varepsilon_n(t) x^{\log n}) = \log(\prod_{d=2}^{\infty} (1-tx^{2\log d})^{-1})$$

$$= \sum_{d=2}^{\infty} \log(1-tx^{2\log d})^{-1}$$

Taking the formal derivative with respect to x, we obtain

$$\frac{d}{dx}(\log(\sum_{n=1}^{\infty} \varepsilon_n(t) x^{\log n})) = \frac{d}{dx}(\sum_{d=2}^{\infty} \log(1-tx^{2\log d})^{-1})$$

$$\frac{d}{dx}(\sum_{n=1}^{\infty} \varepsilon_n(t) x^{\log n}) \left(\frac{1}{\sum_{n=1}^{\infty} \varepsilon_n(t) x^{\log n}} \right) = \frac{d}{dx}(\sum_{d=2}^{\infty} \log(1-tx^{2\log d})^{-1})$$

$$\frac{\sum_{n=1}^{\infty} \varepsilon_n(t) \cdot \log n \cdot x^{\log n -1}}{\sum_{n=1}^{\infty} \varepsilon_n(t) x^{\log n}} = \sum_{d=2}^{\infty} \frac{2t \cdot \log d \cdot x^{2\log d -1} \cdot (1-tx^{2\log d})^{-2}}{(1-tx^{2\log d})^{-1}}$$

$$= \sum_{d=2}^{\infty} 2t \cdot \log d \cdot x^{2\log d -1} \cdot (1-tx^{2\log d})^{-1}$$

$$\frac{\frac{1}{x} \cdot \sum_{n=1}^{\infty} \varepsilon_n(t) \cdot \log n \cdot x^{\log n}}{\sum_{n=1}^{\infty} \varepsilon_n(t) x^{\log n}} = \frac{1}{x} \cdot \sum_{d=2}^{\infty} 2t \cdot \log d \cdot x^{2\log d} \cdot (1-tx^{2\log d})^{-1}$$

Multiplying through by x gives

$$\frac{\sum_{n=1}^{\infty} \varepsilon_n(t) \cdot \log n \cdot x^{\log n}}{\sum_{n=1}^{\infty} \varepsilon_n(t) x^{\log n}} = \sum_{d=2}^{\infty} 2t \cdot \log d \cdot x^{2\log d} \cdot (1-tx^{2\log d})^{-1}$$

$$\sum_{n=1}^{\infty} \varepsilon_n(t) \cdot \log n \cdot x^{\log n} = (\sum_{n=1}^{\infty} \varepsilon_n(t) x^{\log n})(\sum_{d=2}^{\infty} 2t \cdot \log d \cdot x^{2\log d} \cdot (1-tx^{2\log d})^{-1})$$

$$= \sum_{n=1}^{\infty} \sum_{d=2}^{\infty} \varepsilon_n(t) x^{\log n} \cdot 2t \cdot \log d \cdot x^{2\log d} \cdot (1-tx^{2\log d})^{-1}$$

$$\sum_{n=1}^{\infty} \varepsilon_n(t) \cdot \log n \cdot x^{\log n} = \sum_{n=1}^{\infty} \sum_{d=2}^{\infty} \varepsilon_n(t) x^{\log n} \cdot 2t \cdot \log d \cdot x^{2\log d} \cdot \sum_{k=0}^{\infty} t^k x^{2k \log d}$$

$$= \sum_{n=1}^{\infty} \sum_{d=2}^{\infty} \sum_{k=0}^{\infty} \varepsilon_n(t) x^{\log n} \cdot 2t \cdot \log d \cdot x^{2\log d} \cdot t^k x^{2k \log d}$$

$$= \sum_{n=1}^{\infty} \sum_{d=2}^{\infty} \sum_{k=0}^{\infty} \varepsilon_n(t) \cdot 2t^{k+1} \cdot \log d \cdot x^{\log n + \log d^2 + \log d^{2k}}$$

$$= \sum_{n=1}^{\infty} \sum_{d=2}^{\infty} \sum_{k=0}^{\infty} \varepsilon_n(t) \cdot 2t^{k+1} \cdot \log d \cdot x^{\log(n \cdot d^{2k+2})}$$

Next, let $m = n \cdot d^{2k+2}$. We then have

$$\sum_{n=1}^{\infty} \varepsilon_n(t) \cdot \log n \cdot x^{\log n} = \sum_{m=4}^{\infty} \sum_{d=2}^{\infty} \sum_{k=0}^{\infty} \varepsilon_{\frac{m}{d^{2k+2}}}(t) \cdot 2t^{k+1} \cdot \log d \cdot x^{\log m}.$$

On the right-hand-side of this equation, we should really sum over those values of m such that $m = n \cdot d^{2(k+1)}$ with $n \geq 1$, $d \geq 2$, $k \geq 0$. However, this means that $\frac{m}{d^{2k+2}} = n \geq 1$, where n is a an integer. So, if we *define* $\varepsilon_{\frac{m}{d^{2k+2}}}(t) = 0$ whenever $\frac{m}{d^{2k+2}}$ is *not* an integer, then we can sum over all $m \geq 1$. Then the coefficient of $x^{\log m}$ will be zero whenever d^{2k+2} does not divide m (for $d \geq 2$, $k \geq 0$). We should note that if $\frac{m}{d^{2k+2}} = r$ is not an integer, then $\varepsilon_r(t) = 0$. But there may be other values of $\varepsilon_r(t)$ that are also equal to zero, even when r is a positive integer. For example, $\varepsilon_{12}(t) = \varepsilon_3(t) = 0$. Continuing from the last equation, and using the observations just mentioned, we have

$$\sum_{m=1}^{\infty} \varepsilon_m(t) \cdot \log m \cdot x^{\log m} = \sum_{m=1}^{\infty} \sum_{d=2}^{\infty} \sum_{k=0}^{\infty} \varepsilon_{\frac{m}{d^{2k+2}}}(t) \cdot 2t^{k+1} \cdot \log d \cdot x^{\log m}$$

$$\sum_{m=1}^{\infty} (\varepsilon_m(t) \cdot \log m) x^{\log m} = \sum_{m=1}^{\infty} (\sum_{d=2}^{\infty} \sum_{k=0}^{\infty} \varepsilon_{\frac{m}{d^{2k+2}}}(t) \cdot 2t^{k+1} \cdot \log d) x^{\log m}$$

Equating coefficients of $x^{\log m}$ on both sides of the equation gives

$$\varepsilon_m(t) \cdot \log m = \sum_{d=2}^{\infty} \sum_{k=0}^{\infty} \varepsilon_{\frac{m}{d^{2k+2}}}(t) \cdot 2t^{k+1} \cdot \log d.$$

Here, when we equate coefficients of $x^{\log m}$, some terms may have zero coefficients. The only values of m for which $x^{\log m}$ has nonzero coefficients are those where d^{2k+2} divides m for $d \geq 2$, $k \geq 0$. If we let $j = k+1$, we get the final result:

$$\varepsilon_m(t) \cdot \log m = 2 \cdot \sum_{d>1} \sum_{j \geq 1} \varepsilon_{\frac{m}{d^{2j}}}(t) \cdot t^j \cdot \log d. \quad \square$$

Example 6.8. Suppose that we have two *red* balls. Clearly, there is only one way to place the two balls into identical boxes, with no box empty, such that each box has even color replication. The solution is simply to place the two balls into one box. (Two boxes will not work, because if we put one red ball into one box and the other red ball into a second box, then each box will contain only *one* red ball. So, there would not be *even color replication* in each box.) Let us use this simple example to verify that $\varepsilon_{p^2}(t) = t$, for any prime p. To show this, we model the problem by letting the color red be represented by a prime number, say $p = 2$. Then, since we have two red balls, the specification number is $n = p^2 = 2^2 = 4$. There is only one non-unit square divisor of 4, namely $d^{2j} = 2^2 = 4$, with $d = 2$ and $j = 1$. By Theorem 6.7, we have

$$\varepsilon_4(t) \cdot \log 4 = 2 \cdot \sum_{d>1} \sum_{j \geq 1} \varepsilon_{\frac{4}{d^{2j}}}(t) \cdot t^j \cdot \log d$$
$$= 2(\varepsilon_1(t) \cdot t \cdot \log 2).$$

Then, since $\varepsilon_4(t) \cdot \log 4 = 2\varepsilon_4(t) \cdot \log 2$, we have $\varepsilon_4(t) \cdot \log 2 = \varepsilon_1(t) \cdot t \cdot \log 2$, or simply $\varepsilon_4(t) = t \cdot \varepsilon_1(t)$. But $\varepsilon_1(t) = 1$. Therefore, $\varepsilon_4(t) = t$, as expected. $\square$

Example 6.9. Let us verify Lemma 6.6 for the specification number $n = 12$. Here, since 12 is not a perfect square, Lemma 6.6 tells us that $\varepsilon_{12}(t) = 0$. By Theorem 6.7, we have

$\varepsilon_{12}(t) \cdot \log 12 = 2 \cdot \sum_{d>1} \sum_{j \geq 1} \varepsilon_{\frac{12}{d^{2j}}}(t) \cdot t^j \cdot \log d$. The only non-unit square divisor of 12 is $d^2 = 4$, with $d = 2$ and $j = 1$. Hence, we have $\varepsilon_{12}(t) \cdot \log 12 = 2 \cdot \varepsilon_{\frac{12}{2^2}}(t) \cdot t^1 \cdot \log 2$. Using the rules of logarithms, this last equation becomes $\varepsilon_{12}(t) \cdot \log 3 + 2 \cdot \varepsilon_{12}(t) \cdot \log 2 = 2 \cdot \varepsilon_3(t) \cdot t \cdot \log 2$. Equating coefficients of $\log 3$ on both sides of the equation gives $\varepsilon_{12}(t) = 0$, which agrees with Lemma 6.6. What happens if we equate the coefficients of log 2 on both sides of the equation? We get $\varepsilon_{12}(t) = t \cdot \varepsilon_3(t)$. But $\varepsilon_3(t) = 0$, because we cannot place a single ball into boxes such that each box has even color replication. (The prime number 3 represents a single ball, and a single ball cannot be paired!) So, we still get the same result: $\varepsilon_{12}(t) = 0$. □

Lemma 6.10. *Let $g_n(t)$ be the total number of unrestricted distributions of objects, of specification n, into identical boxes with no box empty. Then $\varepsilon_{n^2}(t) = g_n(t)$.*

Proof. Let the specification number be $n = p_1^{\alpha_1} p_2^{\alpha_2} \cdots p_k^{\alpha_k}$. For each distribution of the colored balls into identical boxes, we can imagine "splitting" each colored ball in two equal halves. Doing so, we end up with an even color-occupancy distribution of "half-balls" corresponding to the specification number $n^2 = p_1^{2\alpha_1} p_2^{2\alpha_2} \cdots p_k^{2\alpha_k}$. Conversely, given any even color-occupancy distribution of colored balls corresponding to the specification $n^2 = p_1^{2\alpha_1} p_2^{2\alpha_2} \cdots p_k^{2\alpha_k}$, we can "pair up" like colors of balls in each box and let each pair be a "super-ball." We then have a distribution of super-balls corresponding to $n = p_1^{\alpha_1} p_2^{\alpha_2} \cdots p_k^{\alpha_k}$. Thus, we have established a bijection between $g_n(t)$ and $\varepsilon_{n^2}(t)$. □

Theorem 6.7 and Lemma 6.10 immediately imply the following theorem, which shows how to evaluate $g_n(t)$ in terms of $\varepsilon_{n^2}(t)$:

Theorem 6.11. $g_n(t) \cdot \log n = \sum_{d>1} \sum_{j \geq 1} \varepsilon_{\frac{n^2}{d^{2j}}}(t) \cdot t^j \cdot \log d$.

Proof. If we replace n by n^2 in Theorem 6.7, we get

$$\varepsilon_{n^2}(t) \cdot \log n^2 = 2 \cdot \sum_{d>1} \sum_{j \geq 1} \varepsilon_{\frac{n^2}{d^{2j}}}(t) \cdot t^j \cdot \log d.$$

Since $\varepsilon_{n^2}(t) = g_n(t)$, by Lemma 6.10, and $\log n^2 = 2 \log n$, we obtain

$$g_n(t) \cdot \log n = \sum_{d>1} \sum_{j \geq 1} \varepsilon_{\frac{n^2}{d^{2j}}}(t) \cdot t^j \cdot \log d. \quad \square \tag{6.5}$$

Identical Objects

If all of the colored balls (objects) are identical, then the distribution of the objects into identical boxes is equivalent to the partitions of a positive integer. In particular, if we let $n = p^k$, for prime p, be the specification number for k identical objects, then $\varepsilon_n(t) = \varepsilon_{p^k}(t)$ gives the distribution polynomial for the number of *partitions* of k into *even parts*. If $\pi_e(k,r)$ is the number of partitions of k into precisely r even parts, then $[t^r]\varepsilon_{p^k}(t) = \pi_e(k,r)$, and $\varepsilon_{p^k}(1) = \pi_e(k)$, which is the total number of partitions of k into even parts. For example, the partition of 8 given by $2 + 2 + 4$ is counted by $\pi_e(8,3)$. We now present a new proof of a known theorem, which provides further validation for the theory.

Theorem 6.12. $k \cdot \pi_e(2k) = \sum_{j=0}^{k-1} \pi_e(2j) \cdot \sigma(k-j)$, where $\sigma(n)$ is the sum of positive integral divisors of n. We define $\pi_e(0) = 1$.

Proof. Let $n = p^{2k}$ for prime p. The exponent of p is assumed even because if the exponent is odd, then $\varepsilon_n(t) = 0$. In other words, we cannot distribute an odd number of balls into boxes such that each box has *even* color replication, for then there would be an even number of balls—a contradiction. Next, by Theorem 6.7, we have the following calculations (next page):

$$\varepsilon_n(t) \cdot \log p^{2k} = 2 \cdot \sum_{d>1} \sum_{j\geq 1} \varepsilon_{\frac{p^{2k}}{d^{2j}}}(t) \cdot t^j \cdot \log d$$

$$2k \cdot \varepsilon_n(t) \cdot \log p = 2 \cdot \sum_{r\geq 1} \sum_{j\geq 1} \varepsilon_{\frac{p^{2k}}{(p^r)^{2j}}}(t) \cdot t^j \cdot \log p^r$$

$$k \cdot \varepsilon_n(t) \cdot \log p = \sum_{r\geq 1} \sum_{j\geq 1} \varepsilon_{\frac{p^{2k}}{(p^r)^{2j}}}(t) \cdot t^j \cdot \log p^r$$

$$= \sum_{r\geq 1} \sum_{j\geq 1} \varepsilon_{p^{2k-2rj}}(t) \cdot t^j \cdot r \cdot \log p$$

$$= \sum_{m\geq 1} \sum_{j\geq 1} \varepsilon_{p^{2(k-m)}}(t) \cdot t^j \cdot \frac{m}{j} \cdot \log p, \text{ with } m = rj$$

$$= \sum_{m=1}^{k} \sum_{j\mid m} \varepsilon_{p^{2(k-m)}}(t) \cdot t^j \cdot \frac{m}{j} \cdot \log p, \text{ with } \varepsilon_{p^{-a}}(t) = 0 \text{ for } a>0$$

$$= \sum_{m=1}^{k} \varepsilon_{p^{2(k-m)}}(t) \sum_{j\mid m} \frac{m}{j} \cdot t^j \cdot \log p$$

Equating coefficients of $\log p$ on both sides of the equation gives

$$k \cdot \varepsilon_{p^{2k}}(t) = \sum_{m=1}^{k} \varepsilon_{p^{2(k-m)}}(t) \cdot P_m(t), \text{ where } P_m(t) = \sum_{j\mid m} \frac{m}{j} t^j. \tag{6.6}$$

Note that $P_m(1) = \sum_{j\mid m} \frac{m}{j} = \sum_{j\mid m} j = \sigma(m)$, the sum of the positive integral divisors of m. Let $\pi_e(n)$ denote the number of partitions of n into even parts. Then $\varepsilon_{p^{2k}}(1) = \pi_e(2k)$. If we let $t=1$ and let $m=k-j$ (or $j=k-m$), equation (6.6) becomes

$$k \cdot \pi_e(2k) = \sum_{j=k-1}^{0} \pi_e(2j) \cdot \sigma(k-j), \text{ a backward sum}$$

$$= \sum_{j=0}^{k-1} \pi_e(2j) \cdot \sigma(k-j), \text{ a forward sum.}$$

In this recurrence relation, we define $\pi_e(0) = 1$ to ensure that we obtain the correct answer for the trivial degenerate cases. □

Example 6.13. To illustrate the application of Theorem 6.12, let us calculate the number of partitions of 6 into even parts. Using the recurrence relation from Theorem 6.12, we have

$$3 \cdot \pi_e(6) = \sum_{j=0}^{2} \pi_e(2j) \cdot \sigma(3-j)$$
$$= \pi_e(0) \cdot \sigma(3) + \pi_e(2) \cdot \sigma(2) + \pi_e(4) \cdot \sigma(1)$$
$$= (1)(4) + (1)(3) + (2)(1)$$
$$= 9$$

Solving for $\pi_e(6)$, we obtain $\pi_e(6) = 3$. In fact, the three partitions of 6 into even parts are 6, 4 + 2, and 2 + 2 + 2. □

The next theorem is an application of Theorem 6.11. Later, we will use Theorem 6.14 to prove an interesting identity for the Stirling numbers of the second kind.

Theorem 6.14. $g_{p_1 p_2 \cdots p_k}(t) = t \cdot \sum_{j=0}^{k-1} \binom{k-1}{j} \cdot \varepsilon_{p_0^2 p_1^2 \cdots p_j^2}(t)$, with $p_0 = 1$.

The inclusion of $p_0 = 1$ is just a notational trick to ensure that the formula works properly in degenerate cases. We could exclude the p_0, since it is unnecessary, but some confusion might result when $j = 0$.

Proof. By Theorem 6.11, we have $g_n(t) \cdot \log n = \sum_{d>1} \sum_{j\geq 1} \varepsilon_{\frac{n^2}{d^{2j}}}(t) \cdot t^j \cdot \log d$. Let $n = p_1 p_2 \cdots p_k$ so that $n^2 = p_1^2 p_2^2 \cdots p_k^2$. By direct application of Theorem 6.11, we have

$$g_{p_1 p_2 \cdots p_k}(t) \cdot \log(p_1 p_2 \cdots p_k) = \sum_{\substack{r=1 \\ r \neq \alpha, \beta, \ldots, \gamma}}^{k} \varepsilon_{p_\alpha^2 p_\beta^2 \cdots p_\gamma^2}(t) \cdot t \cdot \log p_r + \sum_{\substack{r,s=1 \\ r \neq s \\ r,s \neq \alpha, \beta, \ldots, \gamma}}^{k} \varepsilon_{p_\alpha^2 p_\beta^2 \cdots p_\gamma^2}(t) \cdot t \cdot \log(p_r p_s) + \ldots$$
$$\ldots + \varepsilon_1(t) \cdot t \cdot \log(p_1 p_2 \cdots p_k)$$

In the first sum on the right side of the equal sign, we are summing over each of the k primes. (That is, we have a term with $\log p_r$ for each of the k primes.) In the second sum, we are summing over $\log(p_r p_s)$ where we are choosing all possible combinations of two distinct primes chosen from k primes. We continue this way until we get to the last term, which is effectively a sum over all combinations of k primes chosen from k primes—of course, this can only be done in one way, so the last term has no actual "summation" symbol. Now, let us equate coefficients of $\log p_1$ on both sides of the last equation. Doing so, we obtain the final result:

$$g_{p_1 p_2 \cdots p_k}(t) = t \cdot \sum_{j=0}^{k-1} \binom{k-1}{j} \cdot \varepsilon_{p_0^2 p_1^2 \cdots p_j^2}(t), \text{ with } p_0 = 1.$$

The binomial coefficients $\binom{n}{r}$ appear in a natural way, because we are summing over all combinations of distinct primes chosen none at a time, one at a time, …, and so on. We also used the fact that specification numbers having the same form have equal distributions. So, for example, $\varepsilon_{p_1^2}(t) = \varepsilon_{p_2^2}(t)$, $\varepsilon_{p_1^2 p_2^2}(t) = \varepsilon_{p_j^2 p_k^2}(t)$, for $j \neq k$, etc. The proof only looks complicated. It is actually quite simple. Try following the proof using a concrete example, such as a case with only three distinct primes. □

Example 6.15. Let us use Theorem 6.14 to calculate the distribution polynomial for $g_{p_1 p_2 p_3}(t)$. By Theorem 6.14 we have

$$\begin{aligned} g_{p_1 p_2 p_3}(t) &= t \cdot \sum_{j=0}^{2} \binom{2}{j} \cdot \varepsilon_{p_0^2 p_1^2 \cdots p_j^2}(t) \\ &= t \cdot (\varepsilon_{p_1^2 p_2^2}(t) + 2\varepsilon_{p_1^2}(t) + \varepsilon_1(t)) \\ &= t \cdot ((t + t^2) + 2t + 1) \\ &= t^3 + 3t^2 + t \quad \Box \end{aligned}$$

We can also use Theorem 6.14 to prove an interesting identity for Stirling numbers of the second kind, which are denoted by $\begin{Bmatrix} k \\ j \end{Bmatrix}$. The Stirling number $\begin{Bmatrix} k \\ j \end{Bmatrix}$ counts the number of ways to partition a set of k distinct objects into j nonempty disjoint subsets.

Theorem 6.16. *For* $k \geq r \geq 1$, $\begin{Bmatrix} k+1 \\ r+1 \end{Bmatrix} = \sum_{j=r}^{k} \binom{k}{j} \begin{Bmatrix} j \\ r \end{Bmatrix}$.

Proof. By Theorem 6.14, we have

$$g_{p_1p_2\cdots p_k}(t) = t \cdot \sum_{j=0}^{k-1} \binom{k-1}{j} \cdot \varepsilon_{p_0^2 p_1^2 \cdots p_j^2}(t), \text{ with } p_0 = 1. \tag{6.7}$$

We know from Chapter 4 that $g_{p_1p_2\cdots p_k}(t) = \sum_{r=1}^{k} \left\{ {k \atop r} \right\} t^r$. Also, since $\varepsilon_{n^2}(t) = g_n(t)$, by Lemma 6.10, we have $\varepsilon_{p_0^2 p_1^2 \cdots p_j^2}(t) = \sum_{m=0}^{j} \left\{ {j \atop m} \right\} t^m$. (Note that $\left\{ {0 \atop 0} \right\} = 1$ and $\left\{ {j \atop 0} \right\} = 0$ for $j > 0$.) Substituting these expressions into (6.7), we obtain

$$\sum_{r=1}^{k} \left\{ {k \atop r} \right\} t^r = t \cdot \sum_{j=0}^{k-1} \binom{k-1}{j} \sum_{m=0}^{j} \left\{ {j \atop m} \right\} t^m$$

$$= t + t \cdot \sum_{j=1}^{k-1} \binom{k-1}{j} \sum_{m=1}^{j} \left\{ {j \atop m} \right\} t^m$$

$$= t + \sum_{j=1}^{k-1} \sum_{m=1}^{j} \binom{k-1}{j} \left\{ {j \atop m} \right\} t^{m+1}.$$

On the right-hand-side, let $r = m+1$. We then get

$$\sum_{r=1}^{k} \left\{ {k \atop r} \right\} t^r = t + \sum_{j=1}^{k-1} \sum_{r=2}^{j+1} \binom{k-1}{j} \left\{ {j \atop r-1} \right\} t^r \tag{6.8}$$

To manipulate the indices of summation, we will make use of the *Iversonian convention*. The Iversonian convention is not well known among most mathematicians, but it is a remarkably simple and powerful way to keep track of, and manipulate, indices of summation. According to the Iversonian convention, $[P(k)] = 1$ if $P(k)$ is true; otherwise $[P(k)] = 0$. This allows us to pull the indices of summation "off to the side" and sum over all possible values of the indices. When we are done manipulating the indices, we can put them back into the summations if desired. Thus, we can write equation (6.8) as

$$\sum_{r=1}^{k} \left\{ {k \atop r} \right\} t^r = t + \sum \binom{k-1}{j} \left\{ {j \atop r-1} \right\} t^r \cdot [k-1 \geq j \geq 1][j+1 \geq r \geq 2]$$

$$= t + \sum \binom{k-1}{j} \left\{ {j \atop r-1} \right\} t^r \cdot [k \geq j+1 \geq 2][j+1 \geq r \geq 2]$$

$$= t + \sum \binom{k-1}{j} \left\{ {j \atop r-1} \right\} t^r \cdot [k \geq j+1 \geq r \geq 2]$$

$$= t + \sum \binom{k-1}{j} \left\{ {j \atop r-1} \right\} t^r \cdot [k \geq r \geq 2][k \geq j+1 \geq r]$$

$$= t + \sum_{r=2}^{k} \sum_{j+1=r}^{k} \binom{k-1}{j} \left\{ {j \atop r-1} \right\} t^r$$

$$= t + \sum_{r=2}^{k} \sum_{j=r-1}^{k-1} \binom{k-1}{j} \left\{ {j \atop r-1} \right\} t^r.$$

If we equate coefficients of t^r on both sides of the equation, we get

$$\left\{ {k \atop r} \right\} = \sum_{j=r-1}^{k-1} \binom{k-1}{j} \left\{ {j \atop r-1} \right\}.$$

By a change of variables, with $r \geq 1$, we can write the last expression as

$$\left\{ {k+1 \atop r+1} \right\} = \sum_{j=r}^{k} \binom{k}{j} \left\{ {j \atop r} \right\}. \quad \square$$

Now let us look at an example, just to verify our result.

Example 6.17. Using a good mathematics reference book, we can look up the Stirling number $\left\{ {4 \atop 2} \right\} = 7$. Using Theorem 6.16 with $k = 3$ and $r = 1$, we get

$$\sum_{j=1}^{3} \binom{3}{j} \left\{ {j \atop 1} \right\} = \binom{3}{1}\left\{ {1 \atop 1} \right\} + \binom{3}{2}\left\{ {2 \atop 1} \right\} + \binom{3}{3}\left\{ {3 \atop 1} \right\}$$
$$= (3)(1) + (3)(1) + (1)(1)$$
$$= 3 + 3 + 1$$
$$= 7.$$

Theorem 6.18. *Let $G(x, t)$ denote the logarithmic generating function for the distribution of arbitrary colored balls (objects) into identical boxes (with no box empty) without occupancy restrictions. Let $\Delta(x, t)$ denote the logarithmic generating function for the distribution of arbitrary colored balls into identical boxes (no box empty) such that the occupancies of the boxes are distinct. Let $E(x, t)$ denote the logarithmic generating function for the distribution of arbitrary colored balls into identical boxes (no box empty) such that each box has even color occupancy. Then*

$$G(x,t) = \Delta(x,t) \cdot E(x,t^2).$$ (6.9)

Proof. By Theorem 6.4, we have

$$E(x,t) = \prod_{d=2}^{\infty} \frac{1}{(1-tx^{2\log d})}$$

$$= \prod_{d=2}^{\infty} \frac{1}{(1-(t^{1/2}x^{\log d})^2)}$$

$$= \prod_{d=2}^{\infty} \frac{1}{(1-t^{1/2}x^{\log d})(1+t^{1/2}x^{\log d})}$$

$$= \prod_{d=2}^{\infty} \frac{1}{(1-t^{1/2}x^{\log d})} \prod_{d=2}^{\infty} \frac{1}{(1+t^{1/2}x^{\log d})}$$

$$= \prod_{d=2}^{\infty} \frac{1}{(1-ux^{\log d})} \prod_{d=2}^{\infty} \frac{1}{(1+ux^{\log d})}, \text{ where } u = t^{1/2}$$

$$= G(x,u) \cdot \frac{1}{\Delta(x,u)}$$

Therefore, we have $G(x,u) = \Delta(x,u) \cdot E(x,t) = \Delta(x,u) \cdot E(x,u^2)$. By change of variable, so that we use t's instead of u's, we have the final result: $G(x,t) = \Delta(x,t) \cdot E(x,t^2)$. □

Corollary 6.19. *Let $g_n(t)$ denote the object distribution polynomial for unrestricted distributions of colored balls of specification n into identical boxes with no box empty. Let $\delta_n(t)$ denote the object distribution polynomial for distributions of colored balls of specification n into identical boxes, with no box empty, such that the occupancies of the boxes are distinct. Let $\varepsilon_n(t)$ denote the object* *distribution polynomial for distributions of colored balls of specification n into identical boxes, with no box empty, such that each box has even color occupancy. Then*

$$g_n(t) = \delta_n(t) * \varepsilon_n(t^2)$$
$$= \sum_{d|n} \delta_d(t) \cdot \varepsilon_{\frac{n}{d}}(t^2). \qquad (6.10)$$

Proof. We know that multiplying two logarithmic generating functions corresponds to taking the convolution product of the terms generated by those functions. Hence, we have

$$G(x,t) = \Delta(x,t) \cdot E(x,t^2)$$

$$\sum_{n=1}^{\infty} g_n(t) x^{\log n} = \sum_{n=1}^{\infty} \delta_n(t) x^{\log n} \cdot \sum_{n=1}^{\infty} \varepsilon_n(t^2) x^{\log n}$$

$$= \sum_{n=1}^{\infty} (\sum_{d|n} \delta_d(t) \cdot \varepsilon_{\frac{n}{d}}(t^2)) x^{\log n}.$$

Equating coefficients of $x^{\log n}$ on both sides of the equation gives the desired result (6.10). □

Corollary 6.20. *Let $g(n)$ denote the total number of ways to distribute colored balls of specification n into identical boxes with no box empty. Let $\delta(n)$ denote the total number of ways to distribute colored balls of specification n into identical boxes with no box empty such that the occupancies of the boxes are distinct. Let $\varepsilon(n)$ denote the total number of ways to distribute colored balls of specification n into identical boxes, with no box empty, such that each box has even color occupancy. Then*

$$g(n) = \delta(n) * \varepsilon(n)$$
$$= \sum_{d|n} \delta(d) \cdot \varepsilon(n/d). \qquad (6.11)$$

Proof. Let $t = 1$ in equation (6.10). □

The result $g(n) = \delta(n) * \varepsilon(n)$ is reminiscent of similar results from number theory, such as $\sigma(n) = \varphi(n) * \tau(n)$. This suggests the existence of combinatorial analogs to number-theory identities.

Remark 6.21. We can interpret Corollary 6.20 in another interesting way. Since the problem of distributing colored balls into identical boxes with no box empty is equivalent to the problem of determining the number of unordered factorizations of a positive integer n, excluding unit factors, we can give Corollary 6.20 a number-theoretic interpretation. Let $g(n)$ denote the number of unordered factorizations of n, excluding unit factors. Let $\delta(n)$ denote the number of unordered factorizations of n into *distinct factors*, excluding unit factors. Let $\varepsilon(n)$ denote the number of unordered factorizations of n into *square factors* (e.g., 4, 9, 16, 25, and so on), excluding unit factors. (Note that $\varepsilon(n) = 0$ if n is not a square.) Then, we have $g(n) = \varepsilon(n) * \delta(n)$, which is an interesting result.

Lemma 6.22. $\varepsilon(n) = 0$ *if n is not a square.*

Proof. Since $\varepsilon(n)$ represents the number of ways to distribute objects of specification n into identical boxes (with no box empty) such that each box contains an even number of objects for each type of object occurring in the box, it follows that such a distribution cannot exist whenever we have an odd number of objects. But an odd number of objects implies that the specification number n has the form $n = p_1^{\alpha_1} p_2^{\alpha_2} \cdots p_k^{\alpha_k}$ for at least one *odd* α_j. This means that n cannot be a square, since a square would have even exponents for all of its prime factors. □

Remark 6.23. In all of the foregoing equations, it is useful to remember that $g(1) = 1$, $\delta(1) = 1$, and $\varepsilon(1) = 1$.

Example 6.24. Let $n = 36$. How many unordered factorizations of 36 exist, excluding unit factors? We will calculate the answer to this question using Corollary 6.20 primarily to demonstrate equation (6.11). By equation (6.11), we have

$$g(36) = \sum_{d|36} \delta(d) \cdot \varepsilon(36/d)$$
$$= \delta(1)\varepsilon(36) + \delta(2)\varepsilon(18) + \delta(3)\varepsilon(12) + \delta(4)\varepsilon(9)$$
$$+ \delta(6)\varepsilon(6) + \delta(9)\varepsilon(4) + \delta(12)\varepsilon(3) + \delta(18)\varepsilon(2)$$
$$+ \delta(36)\varepsilon(1)$$
$$= (1)(2) + (1)(0) + (1)(0) + (1)(1) + (2)(0) + (1)(1)$$
$$+ (3)(0) + (3)(0) + (5)(1)$$
$$= 9.$$

This is correct, since $g(36) = 9$. □

Corollary 6.25. $g(n) = \sum_{d^2|n} \delta(n/d^2) \cdot \varepsilon(d^2)$.

Proof. This result follows immediately from Corollary 6.20 using the fact that convolution is commutative and, by Lemma 6.22, $\varepsilon(m) = 0$ whenever m is not a square. □

Corollary 6.26. $g_n(t) = \sum_{d^2|n} g_d(t^2) \cdot \delta_{\frac{n}{d^2}}(t)$. *Here, we are summing over all the square divisors of n.*

Proof. From Corollary 6.19, and using the fact that the convolution product is commutative, we have
$g_n(t) = \varepsilon_n(t^2) * \delta_n(t) = \sum_{d|n} \varepsilon_d(t^2) \cdot \delta_{\frac{n}{d}}(t)$. Since $\varepsilon_d(t^2) = 0$ if d is not a perfect square, we have
$g_n(t) = \sum_{d^2|n} \varepsilon_{d^2}(t^2) \cdot \delta_{\frac{n}{d^2}}(t)$, since the zero terms vanish. But $\varepsilon_{d^2}(t) = g_d(t)$, or $\varepsilon_{d^2}(t^2) = g_d(t^2)$. Hence, by substitution, we have the final result: $g_n(t) = \sum_{d^2|n} g_d(t^2) \cdot \delta_{\frac{n}{d^2}}(t)$. □

Example 6.27. Let us use Corollary 6.26 to calculate $g_{36}(t)$. We have
$$g_{36}(t) = g_1(t^2) \cdot \delta_{36}(t) + g_2(t^2) \cdot \delta_9(t) + g_3(t^2) \cdot \delta_4(t) + g_6(t^2) \cdot \delta_1(t)$$
$$= (1)(t + 3t^2 + t^3) + (t^2)(t) + (t^2)(t) + (t^2 + t^4)(1)$$
$$= t + 4t^2 + 3t^3 + t^4. \ \square$$

Remark 6.28. The function $\varepsilon_n(t)$ is completely characterized as follows:

$$\varepsilon_n(t) = \begin{cases} 0, & \text{if } n \neq m^2 \text{ for } m \in N; \\ g_m(t), & \text{if } n = m^2 \text{ for } m \in N. \end{cases}$$

Theorem 6.29. *Let $\pi(k)$ denote the number of partitions of a positive integer k. Let $\pi_e(k)$ denote the number of partitions of k into even parts. Let $\pi_d(k)$ denote the number of partitions of k into distinct parts. Then*

$$\pi(k) = \sum_{r=0}^{\lfloor k/2 \rfloor} \pi_e(2r) \cdot \pi_d(k-2r). \tag{6.12}$$

We define $\pi_e(0) = 1$ and $\pi_d(0) = 1$.

Proof. Let $n = p^k$ for p prime. By Corollary 6.20, we have

$$g(p^k) = \sum_{d \mid p^k} \delta(d) \cdot \varepsilon(p^k/d). \tag{6.13}$$

Since $\varepsilon(p^m) = 0$ for m odd, equation (6.13) simplifies to

$$g(p^k) = \sum_{d \text{ even}} \delta(p^{k-d}) \cdot \varepsilon(p^d)$$

$$= \sum_{r=0}^{\lfloor k/2 \rfloor} \delta(p^{k-2r}) \cdot \varepsilon(p^{2r}).$$

But $g(p^k) = \pi(k)$, $\delta(p^{k-2r}) = \pi_d(k-2r)$, and $\varepsilon(p^{2r}) = \pi_e(2r)$. Thus, we have

$$\pi(k) = \sum_{r=0}^{\lfloor k/2 \rfloor} \pi_d(k-2r) \cdot \pi_e(2r). \quad \square$$

Let us stand back and look at the big picture. Note that Theorem 6.29 is really a special case of Corollary 6.20. Theorem 6.29 is a well-known theorem in number theory. Corollary 6.20, which is a generalization of Theorem 6.29, is either a new theorem, or it is a theorem that is not well-known. (I cannot say which, because I discovered Theorem 6.18 (and Corollary 6.20) by myself.) This means that we can view the numbers $g(n)$ as "generalized partitions," and we can view Corollary 6.20 as a generalization of Theorem 6.29. If we make a change of variables and compare the two theorems side-by-side, we readily observe the similarities. This also suggests that similar partition theorems probably have similar generalizations.

Special Case	General Case
Integer Partitions	Generalized Partitions

$$\pi(n) = \sum_{r=0}^{n} \pi_d(r) \cdot \pi_e(n-r) \qquad g(n) = \sum_{r|n} \delta(r) \cdot \varepsilon(n/r)$$

Example 6.30. There are 42 integer partitions of the number 10. Let us calculate this number using Theorem 6.29. By Theorem 6.29, we have

$$\pi(10) = \sum_{r=0}^{5} \pi_e(2r) \cdot \pi_d(10-2r)$$
$$= \pi_e(0) \cdot \pi_d(10) + \pi_e(2) \cdot \pi_d(8) + \pi_e(4) \cdot \pi_d(6) + \pi_e(6) \cdot \pi_d(4) + \pi_e(8) \cdot \pi_d(2) + \pi_e(10) \cdot \pi_d(0)$$
$$= (1)(10) + (1)(6) + (2)(4) + (3)(2) + (5)(1) + (7)(1)$$
$$= 42. \ \square$$

Lemma 6.31. $\varepsilon(p_1^2 p_2^2 \cdots p_k^2) = B(k)$, *where* $B(k)$ *is the kth Bell number. The Bell numbers count the total number of ways to partition a non-empty k-set into non-empty disjoint subsets.*

Proof. From Lemma 6.10, with $t=1$, we have

$$\varepsilon(p_1^2 p_2^2 \cdots p_k^2) = g(p_1 p_2 \cdots p_k)$$
$$= \sum_{r=1}^{k} \begin{Bmatrix} k \\ r \end{Bmatrix}$$
$$= B(k). \ \square$$

For the purposes of calculating the object distribution polynomials $g_n(t)$, it would be very useful if we could easily calculate $g_{n^2}(t)$ in terms of simpler polynomials that we already know. For then if we wanted to calculate $g_{105}(t)$ we could first calculate $g_{10}(t)$, then calculate $g_{10^2}(t) = g_{100}(t)$, and then use other theorems to "step" our way up to $g_{105}(t)$. Although I have not yet found a simple way of accomplishing this feat, the following theorem appears to be a step in that direction.

Theorem 6.32. $g(p_1^2 p_2^2 \cdots p_k^2) = \sum_{r=0}^{k} \binom{k}{r} \cdot \delta(p_0^2 p_1^2 \cdots p_r^2) \cdot B(k-r)$, where $p_0 \equiv 1$, $B(0) \equiv 1$, and $B(n)$ is the nth Bell number.

Proof. Let $n = p_1^2 p_2^2 \cdots p_k^2$. Then, by Corollary 6.20, we have

$$g(n) = \sum_{d \mid n} \delta(d) \cdot \varepsilon(n/d). \tag{6.14}$$

Since $\varepsilon(m) = 0$ if m is not a perfect square, equation (6.14) becomes

$$g(n) = \sum_{d^2 \mid n} \delta(d^2) \cdot \varepsilon(n/d^2). \tag{6.15}$$

In the summation (6.15), $d^2 \in \{1, p_1^2, p_2^2, ..., p_k^2, p_1^2 \cdot p_2^2, p_1^2 \cdot p_3^2, ..., p_1^2 \cdot p_2^2 \cdot p_3^2, ...\}$. In other words, d^2 is the square of a product of distinct primes chosen none at a time, one at a time, two at a time, three at a time, and so on. Since distributions having the same *form* are equal, we can write equation (6.15) as

$$g(n) = \sum_{r=0}^{k} \binom{k}{r} \cdot \delta(p_0^2 p_1^2 \cdots p_r^2) \cdot \varepsilon\left(\frac{n}{p_0^2 p_1^2 \cdots p_r^2}\right). \tag{6.16}$$

Since $n = p_1^2 p_2^2 \cdots p_k^2$, it follows that $\dfrac{n}{p_0^2 p_1^2 \cdots p_r^2}$ is itself a product of the squares of distinct primes.

Hence, by Lemma 6.31, $\varepsilon\left(\dfrac{n}{p_0^2 p_1^2 \cdots p_r^2}\right) = \varepsilon(p_{r+1}^2 p_{r+2}^2 \cdots p_k^2) = \varepsilon(p_1^2 p_2^2 \cdots p_{k-r}^2) = B(k-r)$. Thus, by substitution into (6.16), we get the final result. □

Example 6.33. To illustrate Theorem 6.32, let $n = p_1^2 p_2^2$. Then, by Theorem 6.32, we have

$$g(n) = \sum_{r=0}^{2} \binom{2}{r} \cdot \delta(p_0^2 p_1^2 \cdots p_r^2) \cdot B(2-r)$$

$$= \binom{2}{0} \cdot \delta(p_0^2) \cdot B(2) + \binom{2}{1} \cdot \delta(p_0^2 p_1^2) \cdot B(1) + \binom{2}{2} \cdot \delta(p_0^2 p_1^2 p_2^2) \cdot B(0)$$

$$= \binom{2}{0} \cdot \delta(1) \cdot B(2) + \binom{2}{1} \cdot \delta(p_1^2) \cdot B(1) + \binom{2}{2} \cdot \delta(p_1^2 p_2^2) \cdot B(0)$$

$$= (1)(1)(2) + (2)(1)(1) + (1)(5)(1)$$

$$= 9. \quad \square$$

See Table 5.1 for several δ polynomials.

We can use Theorem 6.32 to derive a "quick and dirty" inequality. The lower bound in the inequality is not very good, but the same could be said of many mathematical inequalities. The real purpose, or application, of mathematical inequalities is to prove other theorems. In other words, we usually do not use inequalities to calculate *numbers*; rather, we use inequalities to prove *theorems*.

Theorem 6.34. $g(p_1^2 p_2^2 \cdots p_k^2) \geq \sum_{r=0}^{k} \binom{k}{r} \cdot B(r) \cdot B(k-r)$.

Proof. If n is the product of distinct primes, then $g(n) = \delta(n)$, because distinct primes represent distinct objects to distribute into boxes. Then, since we have $\delta(p_0^2 p_1^2 \cdots p_r^2) \geq \delta(p_0 p_1 \cdots p_r) = g(p_0 p_1 \cdots p_r) = B(r)$, substitution into Theorem 6.32 immediately gives the desired result. $\square$

Example 6.35. From Example 6.33, we know that $g(p_1^2 p_2^2) = 9$. Using Theorem 6.34, we have

$$g(p_1^2 p_2^2) \geq \sum_{r=0}^{2} \binom{2}{r} \cdot B(r) \cdot B(2-r)$$

$$= \binom{2}{0} \cdot B(0) \cdot B(2) + \binom{2}{1} \cdot B(1) \cdot B(1) + \binom{2}{2} \cdot B(2) \cdot B(0)$$

$$= (1)(1)(2) + (2)(1)(1) + (1)(2)(1)$$

$$= 6.$$

Thus, $g(p_1^2 p_2^2) \geq 6$. $\square$

Theorem 6.36. $G(x, t) \cdot G(x, -t) = E(x, t^2)$.

Proof. By Theorem 5.19, we have $\Delta(x, t) \cdot G(x, -t) = 1$. Note also that $\Delta(x, t) \cdot E(x, t^2) = G(x, t)$:

$$\Delta(x, t) \cdot E(x, t^2) = \prod_{d>1}(1+tx^{\log d}) \cdot \prod_{d>1}\frac{1}{(1-t^2 x^{2\log d})}$$

$$= \prod_{d>1}(1+tx^{\log d}) \cdot \frac{1}{(1+tx^{\log d})(1-tx^{\log d})}$$

$$= \prod_{d>1}\frac{1}{(1-tx^{\log d})}$$

$$= G(x, t)$$

Putting the two equations $\Delta(x, t) \cdot G(x, -t) = 1$ and $\Delta(x, t) \cdot E(x, t^2) = G(x, t)$ together, we get $G(x, t) \cdot G(x, -t) = E(x, t^2)$ □

Corollary 6.37. *For positive integer* $n > 1$, $\sum_{d|n} g(d) \cdot g_{\frac{n}{d}}(-1) = \varepsilon_n(1) = \varepsilon(n)$.

Proof. By Theorem 6.36, we have $G(x, t) \cdot G(x, -t) = E(x, t^2)$. Hence, we have

$$\sum_{n=1}^{\infty} g_n(t) x^{\log n} \cdot \sum_{n=1}^{\infty} g_n(-t) x^{\log n} = \sum_{n=1}^{\infty} \varepsilon_n(t^2) x^{\log n}$$

$$g_n(t) * g_n(-t) = \varepsilon_n(t^2)$$

If we now let $t = 1$, and use the facts that $g_d(1) = g(d)$ and $\varepsilon_n(1) = \varepsilon(n)$, we get

$$\sum_{d|n} g(d) \cdot g_{\frac{n}{d}}(-1) = \varepsilon(n). \quad \square$$

Corollary 6.38. Another form of Corollary 6.37, which follows from the convolution product $g_n(t) * g_n(-t) = \varepsilon_n(t^2)$, is $\sum_{d|n} g_d(t) \cdot g_{\frac{n}{d}}(-t) = \varepsilon_n(t^2)$.

The significance of Corollary 6.38 is that it allows us to calculate the polynomials $\varepsilon_n(t)$ using only the polynomials $g_n(t)$. Example 6.39, below, shows how to do this.

Example 6.39. Let $n = p^2 = 2^2 = 4$. Note that we used the prime number $p = 2$, though any prime can be used. Using a table of values for $g_n(t)$, we note that $g_1(t) = 1$, $g_2(t) = t$, and $g_4(t) = t + t^2$. Then, by Corollary 6.38, $\sum_{d|n} g_d(t) \cdot g_{\frac{n}{d}}(-t) = \varepsilon_n(t^2)$. For $n = 4$, this identity becomes $\sum_{d|4} g_d(t) \cdot g_{\frac{4}{d}}(-t) = \varepsilon_4(t^2)$. Expanding this identity, we get

$$g_1(t) \cdot g_4(-t) + g_2(t) \cdot g_2(-t) + g_4(t) \cdot g_1(-t) = \varepsilon_4(t^2)$$
$$1 \cdot (-t + t^2) + t(-t) + (t + t^2) \cdot 1 = \varepsilon_4(t^2)$$
$$t^2 = \varepsilon_4(t^2).$$

By a change of variable, we immediately obtain $\varepsilon_4(t) = t$. This is, of course, the correct answer. Since $4 = 2^2$, if we let the prime number 2 represent, say, a "red" colored ball, then we can distribute two red balls into one box, with even color occupancy per box, in only one way. And, as expected, the coefficient of t in $\varepsilon_4(t)$ is 1. □

Corollary 6.40. *If $n = p_1 \cdot p_2 \cdot ... \cdot p_k$, for distinct primes p_j, then $\sum_{d|n} g_d(1) \cdot g_{\frac{n}{d}}(-1) = 0$.*

Note that if the specification number n is a product of *distinct* primes, as hypothesized in this corollary, then that means we are distributing a set of *distinct* objects into identical boxes (with no box empty).

Proof. The proof is a simple consequence of Corollary 6.38 with $t = 1$. If the objects we are distributing into boxes are all distinct, then it is impossible to have "even color occupancy" in any box. Therefore, we must have $\varepsilon_n(1) = 0$. □

Theorem 6.41. *Let $\pi(n, t)$ be the partition polynomial for the partitions of n, where n is a positive integer. Let $\pi_{even}(k)$ denote the number of partitions of the positive integer k into even parts. Also, define $\pi(0, 1) = 1$ and $\pi(0, -1) = 1$. We have the following identity:*

$$\sum_{r=0}^{k} \pi(r, 1) \cdot \pi(k-r, -1) = \pi_{even}(k).$$

Proof. By Corollary 6.37, we have $\sum_{d|n} g(d) \cdot g_{\frac{n}{d}}(-1) = \varepsilon_n(1) = \varepsilon(n)$ for positive integer $n > 1$. For the special case where n is a power of a prime, say $n = p^k$, $g(n) = g(p^k) = \pi(k) = \pi(k, 1)$, and $g_{\frac{n}{d}}(-1) = g_{\frac{p^k}{p^r}}(-1) = g_{p^{k-r}}(-1) = \pi(k-r, -1)$. Also, $\varepsilon(n) = \varepsilon(p^k) = \pi_{even}(k)$. Hence, by substitution into the identity of Corollary 6.37, we have $\sum_{r=0}^{k} \pi(r, 1) \cdot \pi(k-r, -1) = \pi_{even}(k)$. □

Example 6.42. Let $k = 6$. Let us demonstrate that $\sum_{r=0}^{6} \pi(r, 1) \cdot \pi(6-r, -1) = \pi_{even}(6)$. We have

$\pi(0, 1)\pi(6, -1) + \pi(1, 1)\pi(5, -1) + \pi(2, 1)\pi(4, -1) + \pi(3, 1)\pi(3, -1) + \pi(4, 1)\pi(2, -1)$
$+ \pi(5, 1)\pi(1, -1) + \pi(6, 1)\pi(0, -1)$
$= (1)(1) + (1)(-1) + (2)(1) + (3)(-1) + (5)(0) + (7)(-1) + (11)(1)$
$= 3.$

By Theorem 6.41, this means that $\pi_{even}(6) = 3$. And that is correct, because there are three partitions of 6 into even parts, namely 6, 2 + 4, and 2 + 2 + 2. In performing the calculations in this example, it is useful to remember that the first several partition polynomials are given by:

$$\pi(1, t) = t$$
$$\pi(2, t) = t + t^2$$
$$\pi(3, t) = t + t^2 + t^3$$
$$\pi(4, t) = t + 2t^2 + t^3 + t^4$$
$$\pi(5, t) = t + 2t^2 + 2t^3 + t^4 + t^5$$
$$\pi(6, t) = t + 3t^2 + 3t^3 + 2t^4 + t^5 + t^6. \quad □$$

A Fundamental Identity for Object Distribution Polynomials: The Square Root Theorem

Generally speaking, when we "multiply" two object distribution polynomials, by taking their convolution product, we do not usually get another object distribution polynomial. So, we don't have a simple closure law for convolution products of object distribution polynomials. However, we do have a kind of weak closure law, which I call the "Square Root Theorem." We call it the Square Root Theorem because, at the risk of abusing notation, we can write it as $g_n(\sqrt{t}) * g_n(-\sqrt{t}) = g_{\sqrt{n}}(t)$. The Square Root Theorem also gives us a fundamental algebraic identity among the general, unrestricted, object distribution polynomials $g_n(t)$. More precisely, we have the following theorem:

Theorem 6.43 (Square Root Theorem). *Let n be a specification number for a set of objects. Let $g_n(t)$ denote the object distribution polynomial for objects of specification n. Then*

$$g_{n^2}(t) * g_{n^2}(-t) = g_n(t^2).$$

Proof. In the proof of Corollary 6.37, we proved that $g_n(t) * g_n(-t) = \varepsilon_n(t^2)$. We also showed that $\varepsilon_{n^2}(t) = g_n(t)$, or $\varepsilon_{n^2}(t^2) = g_n(t^2)$. Then, performing a change of variable by replacing n with n^2, we get

$$g_{n^2}(t) * g_{n^2}(-t) = \varepsilon_{n^2}(t^2)$$
$$= g_n(t^2). \quad \square$$

Example 6.44. Let $n = p_1 p_2$, for distinct primes p_1, p_2. Since $g_{p_1 p_2}(t) = t + t^2$ (Appendix A), we have $g_{p_1 p_2}(t^2) = t^2 + t^4$. On the other hand, we have (next page):

$$g_{p_1^2 p_2^2}(t) * g_{p_1^2 p_2^2}(-t) = \sum_{d | p_1^2 p_2^2} g_d(t) g_{\frac{p_1^2 p_2^2}{d}}(-t)$$

$$= g_1(t) g_{p_1^2 p_2^2}(-t) + g_{p_1}(t) g_{p_1 p_2^2}(-t) + g_{p_2}(t) g_{p_1^2 p_2}(-t) + g_{p_1^2}(t) g_{p_2^2}(-t)$$
$$+ g_{p_2^2}(t) g_{p_1^2}(-t) + g_{p_1 p_2}(t) g_{p_1 p_2}(-t) + g_{p_1^2 p_2}(t) g_{p_2}(-t) + g_{p_1 p_2^2}(t) g_{p_1}(-t)$$
$$+ g_{p_1^2 p_2^2}(t) g_1(-t)$$

$$= 1 \cdot (-t + 4t^2 - 3t^3 + t^4) + t \cdot (-t + 2t^2 - t^3) + t \cdot (-t + 2t^2 - t^3)$$
$$+ (t + t^2)(-t + t^2) + (t + t^2)(-t + t^2) + (t + t^2)(-t + t^2) + (t + 2t^2 + t^3)(-t)$$
$$+ (t + 2t^2 + t^3)(-t) + (t + 4t^2 + 3t^3 + t^4) \cdot 1$$

$$= t^2 + t^4, \text{ as expected.}$$

Theorem 6.43 can be applied to integer partition theory as shown in the following theorem.

Theorem 6.45. *Let $p(n, t)$ be an integer partition polynomial, where the coefficient of t^k is the number of ways to partition the positive integer n into precisely k parts. Then*

$$\sum_{r=0}^{2k} p(r, t) p(2k - r, -t) = p(k, t^2).$$

Proof. By Theorem 6.43, we have $g_{n^2}(t) * g_{n^2}(-t) = g_n(t^2)$. Let $n = p^k$ for a prime p and integer $k \geq 1$. For integer partitions, $g_{p^j}(t) = p(j, t)$. Then we have

$$g_{n^2}(t) * g_{n^2}(-t) = g_{p^{2k}}(t) * g_{p^{2k}}(-t)$$
$$= \sum_{d | p^{2k}} g_d(t) g_{\frac{p^{2k}}{d}}(-t), \text{ where } d = p^r, r \geq 0$$
$$= \sum_{r=0}^{2k} p(r, t) p(2k - r, -t)$$
$$= p(k, t^2). \square$$

Example 6.46. Let $k = 2$. For reference, the relevant integer distribution polynomials are given by

$$p(0, t) = 1$$
$$p(1, t) = t, \text{ namely } \{1\}$$
$$p(2, t) = t + t^2, \text{ namely } \{2, 1+1\}$$
$$p(3, t) = t + t^2 + t^3, \text{ namely } \{3, 1+2, 1+1+1\}$$
$$p(4, t) = t + 2t^2 + t^3 + t^4, \text{ namely } \{4, 1+3, 2+2, 1+1+2, 1+1+1+1\}$$

For $t = 1$ and $k = 2$, Theorem 6.45 implies that $\sum_{r=0}^{4} p(r, 1)p(4-r, -1) = p(2, 1)$, or since

$p(n, 1) = p(n)$, $\sum_{r=0}^{4} p(r)p(4-r, -1) = p(2)$. Let us verify this result. On the right-hand side we have

$p(2) = 2$, since there are 2 integer partitions of the number 2. On the left-hand side we have

$$\sum_{r=0}^{4} p(r)p(4-r, -1) = p(0)p(4, -1) + p(1)p(3, -1) + p(2)p(2, -1) + p(3)p(1, -1) + p(4)p(0, -1)$$
$$= (1)(1) + (1)(-1) + (2)(0) + (3)(-1) + (5)(1)$$
$$= 2$$
$$= p(k), \text{ for } k = 2.$$

Building Even Occupancy Polynomials from Distinct Occupancy Polynomials

We now want to show that we can, at least in principle, use the distinct occupancy object distribution polynomials $\delta_n(t)$ to build object distribution polynomials for even type occupancy $\varepsilon_n(t)$. The following theorem gives the basic relationship. Again, the key point here is that we *can*, at least in principle, use the polynomials $\delta_n(t)$ to build other kinds of distribution polynomials. I am not saying that this is an easy thing to do. I am simply saying that it can be done. In actual practice, we would generally use recurrence relations as a computationally more feasible way to calculate object distribution polynomials.

Lemma 6.47. $G(x, t) = \prod_{k=0}^{\infty} \Delta(x^{2k}, t^{2k})$.

Proof. We prove this result by repeated multiplication as follows:

$$G(x,t) = \prod_{d>1} \frac{1}{(1-tx^{\log d})}$$

$$= \prod_{d>1} \frac{1}{(1-tx^{\log d})} \frac{(1+tx^{\log d})}{(1+tx^{\log d})}$$

$$= \prod_{d>1} \frac{(1+tx^{\log d})}{(1-t^2 x^{2\log d})}$$

$$= \prod_{d>1} \frac{(1+tx^{\log d})}{(1-t^2 x^{2\log d})} \frac{(1+t^2 x^{2\log d})}{(1+t^2 x^{2\log d})}$$

$$= \prod_{d>1} \frac{(1+tx^{\log d})(1+t^2 x^{2\log d})}{(1-t^4 x^{4\log d})}$$

$$= \ldots$$

$$= \prod_{k \geq 0} \Delta(x^{2k}, t^{2k}). \quad \square$$

Theorem 6.48. *Let $E(x,t)$ be the logarithmic generating function for enumerating the distributions of objects into identical boxes, with no box empty, such that each box has even type occupancy. Then the following relation holds true:*

$$E(x,t) = \prod_{k=0}^{\infty} \Delta(x^{2^k}, t^{2^{k-1}}) \cdot \Delta(x^{2^k}, (-1)^{2^k} t^{2^{k-1}}).$$

Proof. By Theorem 6.18, we have the identity $G(x,t) = \Delta(x,t) \cdot E(x,t^2)$. By Theorem 5.19, we also have $\dfrac{1}{\Delta(x,t)} = G(x,-t)$. Putting these relationships together, we get

$$E(x,t^2) = \frac{G(x,t)}{\Delta(x,t)}$$

$$= G(x,t) \cdot G(x,-t)$$

$$= \prod_{k=0}^{\infty} \Delta(x^{2^k}, t^{2^k}) \cdot \prod_{k=0}^{\infty} \Delta(x^{2^k}, (-t)^{2^k})$$

$$= \prod_{k=0}^{\infty} \Delta(x^{2^k}, t^{2^k}) \cdot \Delta(x^{2^k}, (-t)^{2^k}).$$

Making the substitution $t^2 \to t$, we obtain the final result:

$$E(x,t) = \prod_{k=0}^{\infty} \Delta(x^{2^k}, t^{2^{k-1}}) \cdot \Delta(x^{2^k}, (-1)^{2^k} t^{2^{k-1}}). \quad \square$$

Exercises

1. Show that $g(36) = \delta(36) * \varepsilon(36)$. You can use Table 5.1.

2. Use Theorem 6.29 to calculate the number of partitions of 6, or $\pi(6)$. What are these partitions? List them.

3. Use Theorem 6.34 to obtain a lower bound on $g(900)$.

4. Use Corollary 6.40 to show that for $n = 30$ we have $\sum_{d|30} g_d(1) g_{\frac{30}{d}}(-1) = 0$. You can use the polynomials in Appendix A.

5. Verify the Square Root Theorem, Theorem 6.43, for $n = 4$. You can use the polynomials in Appendix A in your calculations.

Chapter 7. Object Distributions Derived from Polynomial Identities

Simple polynomial identities can be used to establish relationships among various object distribution functions. The object distribution functions are logarithmic generating functions that count the number of ways to distribute objects of arbitrary specification into identical boxes, with no box empty, subject to various constraints. By specializing the object distribution identities to cases where all objects are identical, we obtain interesting convolution identities in the theory of integer partitions.

We first make several definitions that will be used in this chapter. The motivation behind the definitions will become apparent when we relate the definitions to a polynomial identity.

Definition 7.1. Let $G(x,t) = \prod_{d>1} \dfrac{1}{(1-tx^{\log d})} = \sum_{n=1}^{\infty} g_n(t) x^{\log n}$.

From Theorem 2.8, we know that $G(x, t)$ is the logarithmic generating function for enumerating the number of ways to distribute objects of specification n into identical boxes with no box empty.

Definition 7.2. Let $T(x,t) = \prod_{d>1} \dfrac{1}{(1-tx^{3\log d})} = \sum_{n=1}^{\infty} T_n(t) x^{\log n}$.

This logarithmic generating function gives us the polynomials $T_n(t)$ for objects of specification n. But what do these polynomials count? If we choose a particular value for d and expand one term of the generating function, we get

$$\dfrac{1}{(1-tx^{3\log d})} = 1 + tx^{3\log d} + t^2 x^{6\log d} + t^3 x^{9\log d} + \ldots$$
$$= 1 + tx^{\log d^3} + t^2 x^{2\log d^3} + t^3 x^{3\log d^3} + \ldots$$

The term $t^k x^{k\log d^3} = (tx^{\log d^3})^k$ represents k identical boxes each having a set of objects (e.g., colored balls) of specification d^3. For each of the k boxes, the colored balls specified by d occur with multiplicity 3. (The "d^3" means that we place *three* complete sets of objects of specification d into each box.) When we expand the product $\prod_{d>1} \frac{1}{(1-tx^{3\log d})}$, over all $d > 1$, we will obtain collections of boxes such that the objects occurring within each box occur in multiples of 3.

To summarize, the distribution polynomial $T_n(t)$ counts the number of distributions of "colored balls" of specification n into non-empty identical boxes such that each box has "triple color multiplicity." In other words, the number of balls of each color within a box is a multiple of 3.

Definition 7.3. Let $M(x,t) = \prod_{d>1} \frac{1}{(1-t^3 x^{3\log d})} = \sum_{n=1}^{\infty} M_n(t) x^{\log n}$.

This logarithmic generating function for $M(x, t)$ appears superficially similar to the logarithmic generating function for $T(x, t)$, but the difference is the "t^3," rather than just "t," in the term $t^3 x^{3\log d}$ (carefully compare Definitions 7.2 and 7.3). What does $M(x, t)$ count? If we choose a particular value of d and expand one term of the generating function, we get

$$\frac{1}{(1-t^3 x^{3\log d})} = 1 + t^3 x^{3\log d} + t^6 x^{6\log d} + t^9 x^{9\log d} + \ldots$$
$$= 1 + t^3 x^{\log d^3} + (t^3)^2 x^{2\log d^3} + (t^3)^3 x^{3\log d^3} + \ldots$$

The term $(t^3)^k x^{k\log d^3} = (tx^{\log d})^{3k}$ represents $3k$ identical boxes where each box contains objects of specification d.

To summarize, the distribution polynomial $M_n(t)$ counts the number of distributions of "colored balls" of specification n into non-empty identical boxes such that the number of *boxes* having some specification $d > 1$ is a multiple of 3. In other words, *the number of boxes* having specification d is 0, or 3, or 6, and so on.

Definition 7.4. Let $D(x,t) = \prod_{d>1} (1 + tx^{\log d} + t^2 x^{2\log d}) = \sum_{n=1}^{\infty} D_n(t) x^{\log n}$.

If we fix a value of $d > 1$, we get the term $1 + tx^{\log d} + t^2 x^{2\log d} = (tx^{\log d})^0 + (tx^{\log d})^1 + (tx^{\log d})^2$. This term counts the number of object distributions of specification d in 0, 1, or 2 boxes. In other words, we have either 0, 1, or 2 boxes such that each box has specification d. When we multiply, or expand, the generating function over all $d > 1$ we get all possible distributions where a given *box* with some specification d occurs with multiplicity 0, 1, or 2.

To summarize, the distribution polynomial $D_n(t)$ counts the number of distributions of "colored balls" (objects) of specification n into non-empty identical boxes such that *at most two boxes* contain identical sets of colored balls.

Relationships Among the Generating Functions

Now that we have defined several logarithmic generating functions for counting various kinds of object distributions into identical boxes, we will establish a few simple relationships among the generating functions. Indeed, the generating functions were defined in order to satisfy a simple polynomial identity. While this approach may seem a bit contrived, later we will see how the relationships can be used to establish non-trivial convolution theorems in the theory of integer partitions.

To establish our relationships among the generating functions, we will make use of the following algebraic identity:

$$(1 - y^3) = (1 - y)(1 + y + y^2). \tag{7.1}$$

A more useful form, for our purposes, is the following:

$$\frac{1}{(1-y)} = \frac{(1 + y + y^2)}{(1 - y^3)}. \tag{7.2}$$

You may be tempted to raise an objection here by claiming that we cannot allow the variable y to take on a value of 1 (since division by zero is undefined). However, for our application to generating functions, convergence is not an issue. We are dealing with a formal algebraic *ring* of power series.

As such, the variable y is nothing more than a placeholder. It does not take on any values, so convergence issues are irrelevant. (We are doing *algebra*, not *analysis*!)

Theorem 7.5. $G(x) = D(x) \cdot T(x)$.

Proof. Since $G(x,t) = \prod_{d>1} \frac{1}{(1-tx^{\log d})} = \sum_{n=1}^{\infty} g_n(t) x^{\log n}$, by Definition 7.1, and setting $t = 1$, we have $G(x) = G(x,1) = \prod_{d>1} \frac{1}{(1-x^{\log d})} = \sum_{n=1}^{\infty} g(n) x^{\log n}$. Similarly, we have $D(x) = D(x, 1) = \prod_{d>1} (1 + x^{\log d} + x^{2\log d}) = \sum_{n=1}^{\infty} D(n) x^{\log n}$ and $T(x) = T(x, 1) = \prod_{d>1} \frac{1}{(1-x^{3\log d})} = \sum_{n=1}^{\infty} T(n) x^{\log n}$. If we let $y(d) = x^{\log d}$, then the theorem immediately follows from the polynomial identity (7.2), because

$$\prod_{d>1} \frac{1}{(1-y(d))} = \prod_{d>1} (1 + y(d) + y(d)^2) \prod_{d>1} \frac{1}{(1-y(d)^3)},$$

or simply $G(x) = D(x) \cdot T(x)$. □

Multiplication of logarithmic generating functions corresponds to *convolution* of the distribution polynomials. That gives us the following corollary to Theorem 7.5:

Corollary 7.6. $g(n) = \sum_{d|n} D(d) \cdot T(n/d)$.

Proof. By Theorem 2.5, multiplication of logarithmic generating functions corresponds to convolution of the distribution polynomials. □

Let us now see how we can apply Corollary 7.6 to the theory of integer partitions. Since the positive integer n specifies the set of objects, say colored balls, what happens if we let $n = p^k$ for some prime p? In that case, all of the colored balls are identical and we have k of them. So, for this special case, distributing the identical balls into identical boxes, with no box empty, is equivalent to the problem of partitioning the positive integer k into a number of parts (the "boxes"). Thus, $g(p^k) = \pi(k)$, $D(p^k) = \pi_D(k)$, and $T(p^k) = \pi_T(k)$, where the "π" functions are partition functions (defined in the statement of the following theorem). Thus, Corollary 7.6 immediately implies the following convolution partition theorem:

Theorem 7.7. *Let $\pi(k)$ denote the number of partitions of the positive integer k into positive integer parts. Let $\pi_D(k)$ denote the number of partitions of k into parts with at most one duplication of each distinct part (i.e., each part occurs at most twice). Let $\pi_T(k)$ denote the number of partitions of k into parts selected from the set $\{3,6,9,12,...\}$ with repetitions allowed. Then we have the following integer partition convolution theorem:*

$$\pi(k) = \sum_{j=0}^{k} \pi_D(j) \cdot \pi_T(k-j)$$
$$= \sum_{j=0}^{k} \pi_T(j) \cdot \pi_D(k-j) \qquad (7.3)$$
$$= \sum_{j \equiv 0 \,(\mathrm{mod}\,3)} \pi_T(j) \cdot \pi_D(k-j),$$

with $\pi(0) = \pi_D(0) = \pi_T(0) = 1$ and $\pi(j) = \pi_D(j) = \pi_T(j) = 0$ for $j < 0$.

Proof. Although partitions are defined for positive integers k, it is convenient to *define* $\pi(0) = \pi_D(0) = \pi_T(0) = 1$, since it simplifies the mathematics. That is because, for distributing objects into boxes, $k = 0$ means that $n = p^k = p^0 = 1$, which *is* allowed in our theory of the logarithmic generating function. We also impose the requirement that $\pi(j) = \pi_D(j) = \pi_T(j) = 0$ for $j < 0$ because we are dealing with partitions. This latter definition allows us to sum over all $j \equiv 0 \,(\mathrm{mod}\,3)$. Setting $n = p^k$, for some prime p, in Corollary 7.6 gives $g(p^k) = \sum_{d|p^k} D(d) \cdot T(p^k/d) = \sum_{j=0}^{k} D(p^j) \cdot T(p^{k-j})$.

Note that every positive integer divisor of p^k has the form p^j for $k \geq j \geq 0$. Distributing *identical* objects into identical boxes is isomorphic to the integer partition problem. Thus, we can rewrite the last expression as $\pi(k) = \sum_{j=0}^{k} \pi_D(j) \cdot \pi_T(k-j) = \sum_{j=0}^{k} \pi_T(j) \cdot \pi_D(k-j)$. Note also that $\pi_T(j) = 0$ unless j is a multiple of 3, for, otherwise, we would not have any partitions of j into parts that are all multiples of 3. Therefore, with the additional constraints given previously, that

$\pi(j) = \pi_D(j) = \pi_T(j) = 0$ for $j < 0$, it is sufficient to carry out the summation over $j \equiv 0 \pmod 3$ with $0 \leq j \leq k$. (This is just a computational short-cut.) □

Note: Before working the examples in this chapter, you may find it useful to first obtain a list of partitions for at least the numbers 1 through 10. Such lists can be found on the Internet, or in a textbook on number theory.

Example 7.8. Let $n = 10$. Since there are 42 partitions of 10, we have $\pi(10) = 42$. From Theorem 7.7, we get the same result:

$$\pi(10) = \sum_{j \equiv 0 (\text{mod } 3)} \pi_T(j) \cdot \pi_D(10 - j)$$
$$= \pi_T(0)\pi_D(10) + \pi_T(3)\pi_D(7) + \pi_T(6)\pi_D(4) + \pi_T(9)\pi_D(1)$$
$$= (1)(22) + (1)(9) + (2)(4) + (3)(1)$$
$$= 42.$$

Theorem 7.9. $G(x, t) = D(x, t) \cdot M(x, t)$.

Proof. This result follows immediately from the polynomial identity (7.2) and the definitions of $G(x, t)$, $D(x, t)$, and $M(x, t)$, since

$$G(x,t) = \prod_{d>1} \frac{1}{(1 - tx^{\log d})}$$
$$= \prod_{d>1}(1 + tx^{\log d} + t^2 x^{2\log d}) \cdot \prod_{d>1} \frac{1}{(1 - t^3 x^{3\log d})}. \quad \square$$

Corollary 7.10. $g_n(t) = \sum_{d|n} D_d(t) \cdot M_{\frac{n}{d}}(t)$.

Proof. By Theorem 2.5, multiplication of logarithmic generating functions corresponds to convolution of the distribution polynomials. □

Corollary 7.11. $g(n) = \sum_{d|n} D(d) \cdot M(n/d)$.

Proof. Set $t = 1$ in Corollary 7.11. □

Theorem 7.12. *Let $\pi(k, t)$ denote the polynomial for the number of partitions of the positive integer k into positive integer parts. The coefficient of t^r in the polynomial $\pi(k, t)$ is the number of partitions of k into precisely r parts. Similarly, let $\pi_D(k,t)$ denote the polynomial for the number of partitions of k into parts with at most one duplication of each distinct part (i.e., each part occurs at most twice). Let $\pi_M(k,t)$ denote the polynomial for the number of partitions of k into parts such that each part occurs with multiplicity 3. Then we have the following integer partition convolution theorem:*

$$\pi(k,t) = \sum_{j=0}^{k} \pi_D(j,t) \cdot \pi_M(k-j,t)$$

$$= \sum_{j=0}^{k} \pi_M(j,t) \cdot \pi_D(k-j,t) \qquad (7.4)$$

$$= \sum_{j \equiv 0 (\bmod 3)} \pi_M(j,t) \cdot \pi_D(k-j,t),$$

with $\pi(0) = \pi_D(0) = \pi_M(0) = 1$ and $\pi(j) = \pi_D(j) = \pi_M(j) = 0$ for $j < 0$.

Proof. The proof is analogous to the proof of Theorem 7.7. Distributing identical objects into identical boxes is isomorphic to the integer partition problem. So, if we let $n = p^k$, for some prime p, in Corollary 7.10, we get the result of Theorem 7.12. □

Corollary 7.13. $\pi(k) = \sum_{j=0}^{k} \pi_D(j) \cdot \pi_M(k-j) = \sum_{j=0}^{k} \pi_M(j) \cdot \pi_D(k-j) = \sum_{j \equiv 0 (\bmod 3)} \pi_M(j) \cdot \pi_D(k-j).$

Proof. Let $t = 1$ in Theorem 7.12. □

Example 7.14. As in Example 7.8, let us examine the partitions of the number 10. We have the following calculations (next page):

$$\pi(10) = \sum_{j \equiv 0 (\bmod 3)} \pi_M(j) \cdot \pi_D(10-j)$$
$$= \pi_M(0) \cdot \pi_D(10) + \pi_M(3) \cdot \pi_D(7) + \pi_M(6) \cdot \pi_D(4) + \pi_M(9) \cdot \pi_D(1)$$
$$= (1)(22) + (1)(9) + (2)(4) + (3)(1)$$
$$= 22 + 9 + 8 + 3$$
$$= 42.$$

Theorem 7.15. *Let $\pi_T(j)$ denote the number of partitions of j into parts selected from the set $\{3,6,9,12,...\}$ with repetitions allowed. In other words, $\pi_T(j)$ is defined as $\pi_T(j) = \pi(j \mid parts \in \{3,6,9,...\})$. Let $\pi_M(j)$ denote the number of partitions of j such that each part occurs with multiplicity 3. In other words, $\pi_M(j)$ is defined as $\pi_M(j) = \pi(j \mid each\ distinct\ part\ occurs\ 3\ times, or\ 6\ times, or\ 9\ times, or\ ...)$. Then $\pi_T(j) = \pi_M(j)$ for all $j \geq 0$.*

Proof. Equating $\pi(k)$ in Theorem 7.7 with Corollary 7.13 gives

$$\sum_{j=0}^{k} \pi_D(j) \cdot \pi_T(k-j) = \sum_{j=0}^{k} \pi_D(j) \cdot \pi_M(k-j).$$ Thus, we have

$$\sum_{j=0}^{k} \pi_D(j) \cdot \pi_T(k-j) - \sum_{j=0}^{k} \pi_D(j) \cdot \pi_M(k-j) = 0$$

$$\sum_{j=0}^{k} \pi_D(j)[\pi_T(k-j) - \pi_M(k-j)] = 0$$

$$\sum_{j=0}^{k} \pi_D(j) \cdot \Delta(k-j) = 0. \ *$$

For each partition of j into parts of multiplicity 3, we can combine each set of the identical parts into a single "part" that is a multiple of 3. Thus, we have $\pi_M(j) \leq \pi_T(j)$. This means that $\Delta(j) \geq 0$ for all $j \geq 0$. Combining this with the fact that $\pi_D(j) > 0$ for all $j \geq 0$, we conclude (from *) that $\Delta(j) = 0$ for all $j \geq 0$. Therefore, $\pi_T(j) = \pi_M(j)$ for all $j \geq 0$. □

Example 7.16. Let $j = 9$. Theorem 7.15 states that $\pi_T(9) = \pi_M(9)$. To determine $\pi_T(9)$ we count the partitions of 9 whose parts are chosen from the set $\{3, 6, 9, \ldots\}$. Those partitions are $3 + 3 + 3$, and $6 + 3$, and 9. Thus, we have $\pi_T(9) = 3$. To determine $\pi_M(9)$ we count the partitions of 9 such that each distinct part occurs with multiplicity three. Those partitions are $1 + 1 + 1 + 1 + 1 + 1 + 1 + 1 + 1$, and $2 + 2 + 2 + 1 + 1 + 1$, and $3 + 3 + 3$. Here, the number of times that each distinct part occurs is a multiple of 3. Thus, we also have $\pi_M(9) = 3$. Therefore, we have $\pi_T(9) = \pi_M(9)$, as expected.

A General Theorem on Integer Partitions

Theorems 7.5 and 7.9, and their associated corollaries, are actually special cases of a more general theorem. First, we make the following definitions:

Definition 7.17. Let $\mu_m(n)$ count the number of distributions of colored balls (objects) of specification n into non-empty identical boxes such that each box has color multiplicity of m. That is, each color that occurs within a given box occurs with multiplicity m. We define a logarithmic generating function for $\mu_m(n)$ as

$$\mu(x) = \sum_{n=1}^{\infty} \mu_m(n) x^{\log n}. \tag{7.5}$$

Definition 7.18. Let $\rho_m(n)$ count the number of distributions of colored balls (objects) of specification n into non-empty identical boxes such that *at most* $m-1$ boxes contain identical sets of colored balls. (Equivalently, no box and its contents occur m or more times.) We define a logarithmic generating function for $\rho_m(n)$ as

$$\rho(x) = \sum_{n=1}^{\infty} \rho_m(n) x^{\log n}. \tag{7.6}$$

Lemma 7.19. $\mu(x) = \prod_{d>1} \dfrac{1}{(1 - x^{m \log d})}$.

Proof. For fixed positive integer $d > 1$, we have

$$\frac{1}{(1-tx^{m\log d})} = 1 + tx^{m\log d} + t^2 x^{2m\log d} + t^3 x^{3m\log d} + \ldots$$
$$= 1 + (tx^{\log d^m}) + (tx^{\log d^m})^2 + (tx^{\log d^m})^3 + \ldots$$

We are using the variable t as a "counter" to count the number of boxes. Thus, the term $(tx^{\log d^m})^k$ represents k boxes such that each of the k boxes contains objects of specification d^m. So, whatever set of colored balls is represented by specification d, there will be m copies of that set in each of the k boxes. When we take the product over all integers $d > 1$, and gather like terms, the coefficient of $x^{\log n}$ will be the number of distributions of colored balls of specification n into (non-empty) identical boxes such that each color within a box occurs a multiple of m times. That number is $\mu_m(n)$, by definition. Finally, let $t = 1$ to get Lemma 7.19. □

Lemma 7.20. $\rho(x) = \prod_{d>1}(1 + x^{\log d} + x^{2\log d} + x^{3\log d} + \ldots + x^{(m-1)\log d})$.

Proof. If we fix a value of $d > 1$, we get the term $1 + x^{\log d} + x^{2\log d} + x^{3\log d} + \ldots + x^{(m-1)\log d} = (x^{\log d})^0 + (x^{\log d})^1 + (x^{\log d})^2 + \ldots + (x^{\log d})^{m-1}$. This term "counts" the number of object distributions of specification d in $0, 1, 2, \ldots,$ or $m-1$ boxes. In other words, we have either $0, 1, 2, \ldots,$ or $m-1$ boxes such that each box has specification d. When we multiply, or expand, the generating function over all $d > 1$ we get all possible distributions where a given box with some specification d occurs with multiplicity $0, 1, 2, \ldots,$ or $m-1$. □

Theorem 7.21. $G(x) = \mu(x) \cdot \rho(x)$.

Proof. Since $(1-y)(1 + y + y^2 + \ldots + y^{m-1}) = 1 - y^m$, or $\dfrac{1}{1-y} = \dfrac{1 + y + y^2 + \ldots + y^{m-1}}{1 - y^m}$, we have

$$\prod_{d>1}\frac{1}{(1-x^{\log d})} = \prod_{d>1}\frac{1}{(1-x^{m\log d})} \cdot \prod_{d>1}(1 + x^{\log d} + x^{2\log d} + \ldots + x^{(m-1)\log d}),$$ which implies that $G(x) = \mu(x) \cdot \rho(x)$ by Lemmas 7.19 and 7.20. □

Corollary 7.22. $g(n) = \sum_{d|n} \mu(d) \cdot \rho(n/d)$.

Proof. The result follows immediately from Theorem 7.21 and the convolution theorem, Theorem 2.5. □

The next result is really a corollary of Theorem 7.21, using Corollary 7.22. However, the result is significant enough that we will state it as a theorem. Theorem 7.23 is an interesting theorem about partitions of a positive integer.

Theorem 7.23. *Let $\pi(k)$ denote the number of partitions of k. Let $\pi_{\mu(m)}(k) = \pi(k \mid parts \in \{m, 2m, 3m, ...\})$. Let $\pi_{\rho(m)}(k) = \pi(k \mid no\ part\ occurs\ more\ than\ m-1\ times) = \pi(k \mid no\ part\ occurs\ m\ or\ more\ times)$. Then the following convolution is true:*

$$\pi(k) = \sum_{j \equiv 0 (\bmod m)} \pi_{\mu(m)}(j) \cdot \pi_{\rho(m)}(k-j). \tag{7.7}$$

Proof. Let $n = p^k$, for prime p, in Corollary 7.22. □

Example 7.24. Let $k = 6$. We know that $\pi(6) = 11$ since there are 11 partitions of the number 6. Let $m = 3$. By Theorem 7.23, we have

$$\begin{aligned}\pi(6) &= \sum_{j \equiv 0 (\bmod 3)} \pi_{\mu(3)}(j) \cdot \pi_{\rho(3)}(6-j) \\ &= \pi_{\mu(3)}(0) \cdot \pi_{\rho(3)}(6) + \pi_{\mu(3)}(3) \cdot \pi_{\rho(3)}(3) + \pi_{\mu(3)}(6) \cdot \pi_{\rho(3)}(0) \\ &= (1)(7) + (1)(2) + (2)(1) \\ &= 7 + 2 + 2 \\ &= 11.\end{aligned}$$

To amplify the terms in this equation, first note that, by definition, $\pi_{\mu(3)}(0) = 1$ and $\pi_{\rho(3)}(0) = 1$. Next, we have $\pi_{\rho(3)}(6) = 7$ because there are 7 partitions of the number 6 where no part occurs three or more times. These are the partitions 2 + 2 + 1 + 1, and 3 + 2 + 1, and 3 + 3, and 4 + 1 + 1, and 4 + 2, and

5 + 1, and 6. Similarly, $\pi_{\rho(3)}(3) = 2$ because there are two partitions of the number 3 where no part occurs three or more times. These are the partitions 2 + 1 and 3. What about, say, $\pi_{\mu(3)}(6)$? We have $\pi_{\mu(3)}(6) = 2$ because there are 2 partitions of the number 6 where the parts are chosen from the set $\{3, 6, 9, ...\}$, namely the partitions 3 + 3 and 6.

Note that Theorem 7.23 is true for any positive integer k and *any* positive integer m.

All Parts Divisible by m and No Parts Divisible by m

Theorem 7.23 can be written in a more spectacular form. Andrews [4], p. 10, states the theorem that $\pi(k \mid no\ part\ divisible\ by\ m) = \pi(k \mid fewer\ than\ m\ copies\ of\ each\ part) = \pi_{\rho(m)}(k)$, using my notations. Thus, by substitution into Theorem 7.23, we can restate Theorem 7.23 in the following elegant form:

Theorem 7.25.
$$\pi(k) = \sum_{j \equiv 0 (\bmod m)} \pi(j \mid all\ parts\ are\ divisible\ by\ m) \cdot \pi(k - j \mid no\ parts\ are\ divisible\ by\ m)$$

Note that the summation is taken over all nonnegative integers that are congruent to zero modulo m, and $j \leq k$.

Corollary 7.26. *For the special case of $m = 2$, Theorem 7.25 becomes*
$$\pi(k) = \sum_{j \equiv 0 (\bmod 2)} \pi(j \mid all\ parts\ are\ even) \cdot \pi(k - j \mid all\ parts\ are\ odd).$$

Example 7.27. Let $m = 2$ and $k = 9$. By Corollary 7.26, we have the following calculations (next page):

$$\pi(9) = \sum_{j \equiv 0 (\bmod 2)} \pi(j \,|\, even\ parts) \cdot \pi(k-j \,|\, odd\ parts)$$

$$= \pi(0\,|\,even\ parts) \cdot \pi(9\,|\,odd\ parts) + \pi(2\,|\,even\ parts) \cdot \pi(7\,|\,odd\ parts)$$
$$+ \pi(4\,|\,even\ parts) \cdot \pi(5\,|\,odd\ parts) + \pi(6\,|\,even\ parts) \cdot \pi(3\,|\,odd\ parts)$$
$$+ \pi(8\,|\,even\ parts) \cdot \pi(1\,|\,odd\ parts)$$
$$= (1)(8) + (1)(5) + (2)(3) + (3)(2) + (5)(1)$$
$$= 8 + 5 + 6 + 6 + 5$$
$$= 30.$$

In fact, there are 30 partitions of 9 as expected. Note that when we sum over all j such that $j \equiv 0 (\bmod 2)$, $\pi(r \,|\, some\ condition\ x) = 0$ for $r < 0$. This restricts the sum to a finite number of terms.

39

Exercises

1. Use Theorem 7.7 to calculate $\pi(8)$, the number of partitions of 8.
2. Use Theorem 7.23 to calculate $\pi(8)$.
3. Use Theorem 7.25 to calculate $\pi(8)$ for $m = 2, 3,$ and 4. (For $m = 2$, Theorem 7.25 becomes Corollary 7.26.)

Chapter 8. Object Distributions with Symmetric Object Sets

Object distributions can be characterized by the objects, the boxes, or the occupancy constraints. We now consider the distribution of a special set of objects into identical boxes with no box empty. The special set of objects that we consider is called a *symmetric object set*. We can think of the objects as colored balls, with repetitions of colors allowed. The set—technically a multiset—of colored balls is *symmetric* if and only if the set is invariant under any permutation of the color labels.

Definition 8.1. A multiset of objects, such as colored balls, is *symmetric* if it is invariant under any permutation of the object labels, or colors.

Example 8.2. Let S be the multiset of colored balls given by $S = \{red, red, white, white, blue, blue\}$, or simply $S = \{r, r, w, w, b, b\}$. The multiset S is symmetric since it is invariant under any permutation of the color labels on the objects. For example, if we apply the permutation, using Cauchy's notation, $\begin{pmatrix} r & w & b \\ b & r & w \end{pmatrix}$ to S, then we get the set of objects $S' = \{b, b, r, r, w, w\}$. But this is just a reordering of the objects in S, so $S' = S$. It's the same set of colored balls, only the colored balls are in a different order. Since the order does not matter, we see that the set S is invariant to this color permutation. You can easily verify that S is invariant to any color permutation. If we assign a unique prime number to each color of object, such as $r = 2$, $w = 3$, and $b = 5$, then S becomes $S = \{2, 2, 3, 3, 5, 5\}$ and a specification number for S is then $n = 2^2 \cdot 3^2 \cdot 5^2 = p_1^2 p_2^2 p_3^2$, where p_1, p_2, and p_3 are distinct primes.

We will now demonstrate that a symmetric multiset of objects has a special form, and this special form is especially apparent when we look at a specification number that represents the multiset of objects.

Theorem 8.3. *A multiset of objects is symmetric if and only if its specification number has the form $n = p_1^m p_2^m \ldots p_k^m$ for distinct primes $p_1, p_2, \ldots, p_k$ and positive integer m.*

Proof. If the specification number for the set of objects has the form $n = p_1^m p_2^m \ldots p_k^m$ for distinct primes $p_1, p_2, \ldots, p_k$, then any permutation of the primes, which are the "color labels" of the objects, simply rearranges the order of the primes in the prime factorization of *n*. So the multiset of objects, which is represented by the prime factors of *n*, does not change. Therefore, the object set is symmetric. Conversely, suppose that the object set *S* is symmetric. When we represent each object type, or color, by a distinct prime number, the set *S* will have a specification number of the form $n = p_1^a p_2^b \ldots p_k^r$. Since *S* is symmetric, by hypothesis, any permutation of the prime factors, or "colors," of *n* leaves *n* unchanged. Suppose, for example, that *n* has the form $n = p_1^a p_2^b q$, where *q* may be composite, and the permutation interchanges, say, p_1 and p_2. Then we have $n = p_1^a p_2^b q = p_2^a p_1^b q = p_1^b p_2^a q$. This means that $p_1^a p_2^b = p_1^b p_2^a$. But, by the Fundamental Theorem of Arithmetic, the prime factorization of *n* is unique. So we must have $a = b$. The same argument applies to any transposition of primes in *n*. And since any permutation can be represented as a product of transpositions, *n* must have the form $n = p_1^m p_2^m \ldots p_k^m$ for some positive integer *m*. □

Our next objective is to derive a recurrence relation for calculating object distribution polynomials having the form $g_{nq}(t)$, where $n = p_1^m p_2^m \cdots p_k^m = (p_1 p_2 \cdots p_k)^m$ for distinct primes $p_1, p_2, \ldots, p_k, q$. The recurrence relation is presented in Theorem 8.6.

Lemma 8.4. *For $(n, q) = 1$ (i.e., gcd(n, q) = 1) and q prime, we have*

$$g_n(t) = \frac{1}{t} \sum_{d|n} \mu(n/d) g_{dq}(t),$$

where $\mu(n)$ is the Möbius function defined by

$$\mu(n) = \begin{cases} 1, & n = 1, \\ (-1)^k, & n = p_1 p_2 \cdots p_k, \text{ distinct primes}, \\ 0, & \text{if } p^2 \text{ divides } n, \text{ prime } p. \end{cases}$$

Proof. By Theorem 4.2, for $(n, q) = 1$, and prime q, we have

$$g_{nq}(t) = t \cdot \sum_{d|n} g_d(t).$$

By the Möbius Inversion Formula, if $f_n(t) = \sum_{d|n} g_d(t)$, then $g_n(t) = \sum_{d|n} \mu(d) f_{\frac{n}{d}}(t) = \sum_{d|n} \mu(n/d) f_d(t)$.

Now let $f_n(t) = \sum_{d|n} g_d(t) = \frac{1}{t} g_{nq}(t)$. Then, using the Möbius Inversion Formula we have

$$g_n(t) = \sum_{d|n} \mu(n/d) f_d(t)$$

$$= \sum_{d|n} \mu(n/d) \frac{1}{t} g_{dq}(t). \quad \square$$

Note that $\mu(n) = 0$ if p^2 divides n for a prime p. We can use this fact to simplify the computation of certain distribution polynomials, since the Möbius function will force many of the terms to zero. To see how this works, consider the following example:

Example 8.5. Let $(n, q) = 1$ for q prime. Let $n = p_1^2 p_2^2$ for distinct primes p_1 and p_2. By Lemma 8.4, we have

$$t \cdot g_{p_1^2 p_2^2}(t) = \sum_{d | p_1^2 p_2^2} \mu(p_1^2 p_2^2 / d) g_{dq}(t)$$

$$= \mu(p_1^2 p_2^2) g_q(t) + \mu(p_1 p_2^2) g_{p_1 q}(t) + \mu(p_1^2 p_2) g_{p_2 q}(t) + \mu(p_2^2) g_{p_1^2 q}(t)$$

$$+ \mu(p_1^2) g_{p_2^2 q}(t) + \mu(p_1) g_{p_1 p_2^2 q}(t) + \mu(p_2) g_{p_1^2 p_2 q}(t) + \mu(p_1 p_2) g_{p_1 p_2 q}(t)$$

$$+ \mu(1) g_{p_1^2 p_2^2 q}(t)$$

$$= \mu(p_1) g_{p_1 p_2^2 q}(t) + \mu(p_2) g_{p_1^2 p_2 q}(t) + \mu(p_1 p_2) g_{p_1 p_2 q}(t) + \mu(1) g_{p_1^2 p_2^2 q}(t).$$

Notice that the Möbius function has eliminated five of nine terms. Rearranging the last expression, we get

$$g_{p_1^2 p_2^2 q}(t) = t \cdot g_{p_1^2 p_2^2}(t) - \mu(p_1) g_{p_1 p_2^2 q}(t) - \mu(p_2) g_{p_1^2 p_2 q}(t) - \mu(p_1 p_2) g_{p_1 p_2 q}(t)$$

$$= t \cdot g_{p_1^2 p_2^2}(t) - (-1)^1 g_{p_1 p_2^2 q}(t) - (-1)^1 g_{p_1^2 p_2 q}(t) - (-1)^2 g_{p_1 p_2 q}(t)$$

$$= t \cdot g_{p_1^2 p_2^2}(t) + (-1)^2 g_{p_1 p_2^2 q}(t) + (-1)^2 g_{p_1^2 p_2 q}(t) + (-1)^3 g_{p_1 p_2 q}(t).$$

Next, note that $g_{p_1 p_2^2 q}(t)$ and $g_{p_1^2 p_2 q}(t)$ have the same "form." Therefore, $g_{p_1 p_2^2 q}(t) = g_{p_1^2 p_2 q}(t)$. We also see that we have terms with $\mu(p_1)$, $\mu(p_2)$, and $\mu(p_1 p_2)$. These terms represent the combinations of one prime selected from the set $\{p_1, p_2\}$ and two primes selected from the set $\{p_1, p_2\}$. So, we can write the preceding equation as

$$g_{p_1^2 p_2^2 q}(t) = t \cdot g_{p_1^2 p_2^2}(t) + \sum_{r=1}^{2} (-1)^{r+1} \binom{2}{r} g_{\frac{p_1^2 p_2^2 q}{p_1 \cdots p_r}}(t).$$

We can generalize this result into the following theorem:

Theorem 8.6. *Let* $n = p_1^m p_2^m \cdots p_k^m$ *for distinct primes* $p_1, p_2, \ldots, p_k$. *Let* q *be a prime such that* $(n, q) = 1$. *Then*

$$g_{nq}(t) = t \cdot g_n(t) + \sum_{r=1}^{k} (-1)^{r+1} \binom{k}{r} g_{\frac{nq}{p_1 p_2 \cdots p_r}}(t).$$

Proof. Let $n = p_1^m p_2^m \cdots p_k^m$ for distinct primes $p_1, p_2, \ldots, p_k$. Let q be a prime such that $(n, q) = 1$. By Lemma 8.4, we have

$$t \cdot g_n(t) = \sum_{d \mid n} \mu(n/d) g_{dq}(t). \tag{8.1}$$

However, by the definition of the Möbius function $\mu(n/d) = 0$ whenever n/d is divisible by the square of a prime, say p_i^2. So the only terms that remain in the summation (8.1) are terms where $\mu(n/d) = \mu(1)$, $\mu(p_i)$, $\mu(p_i p_j)$, ..., or $\mu(p_i p_j \cdots p_k)$, for distinct primes $p_i, p_j, \ldots, p_k$. Thus, (8.1) simplifies to

$$t \cdot g_n(t) = \mu(1) g_{nq}(t) + (\mu(p_1) g_{\frac{nq}{p_1}}(t) + \mu(p_2) g_{\frac{nq}{p_2}}(t) + \ldots + \mu(p_k) g_{\frac{nq}{p_k}}(t))$$
$$+ (\mu(p_1 p_2) g_{\frac{nq}{p_1 p_2}}(t) + \mu(p_1 p_3) g_{\frac{nq}{p_1 p_3}}(t) + \ldots + \mu(p_{k-1} p_k) g_{\frac{nq}{p_{k-1} p_k}}(t))$$
$$+ \ldots + \mu(p_1 p_2 \cdots p_k) g_{\frac{nq}{p_1 p_2 \cdots p_k}}(t).$$

Since the polynomials within each pair of parentheses have the same form, they are equal. For example, $g_{\frac{nq}{p_1}}(t) = g_{\frac{nq}{p_2}}(t) = \ldots = g_{\frac{nq}{p_k}}(t)$ and $g_{\frac{nq}{p_1 p_2}}(t) = g_{\frac{nq}{p_2 p_3}}(t) = \ldots = g_{\frac{nq}{p_{k-1} p_k}}(t)$. Also, we have

$\mu(p_1) = \mu(p_2) = \ldots = \mu(p_k)$ and $\mu(p_1 p_2) = \mu(p_1 p_3) = \ldots = \mu(p_{k-1} p_k)$, and so on. Hence, by counting the number of occurrences of each term having the same form, we have

$$t \cdot g_n(t) = \mu(1) g_{nq}(t) + \binom{k}{1} \mu(p_1) g_{\frac{nq}{p_1}}(t) + \binom{k}{2} \mu(p_1 p_2) g_{\frac{nq}{p_1 p_2}}(t) + \ldots + \binom{k}{k} \mu(p_1 p_2 \cdots p_k) g_{\frac{nq}{p_1 p_2 \cdots p_k}}(t)$$

$$= g_{nq}(t) + \binom{k}{1}(-1)^1 g_{\frac{nq}{p_1}}(t) + \binom{k}{2}(-1)^2 g_{\frac{nq}{p_1 p_2}}(t) + \ldots + \binom{k}{k}(-1)^k g_{\frac{nq}{p_1 p_2 \cdots p_k}}(t)$$

$$= g_{nq}(t) + \sum_{r=1}^{k} (-1)^r \binom{k}{r} g_{\frac{nq}{p_1 p_2 \cdots p_r}}(t).$$

Solving for $g_{nq}(t)$, we get

$$g_{nq}(t) = t \cdot g_n(t) - \sum_{r=1}^{k} (-1)^r \binom{k}{r} g_{\frac{nq}{p_1 p_2 \cdots p_r}}(t)$$

$$= t \cdot g_n(t) + \sum_{r=1}^{k} (-1)^{r+1} \binom{k}{r} g_{\frac{nq}{p_1 p_2 \cdots p_r}}(t). \quad \square$$

Theorem 8.6 is another tool in our arsenal that allows us to build general occupancy object distribution polynomials recursively. We demonstrate how to use Theorem 8.6 in the next two examples.

Example 8.7. Let $n = p_1 p_2 p_3 p_4$. Find $g_{p_1 p_2 p_3 p_4 p_5}(t)$ where $(n, p_5) = 1$.

Solution. By Theorem 8.6, and using polynomials from Appendix A, Table 0, we have

$$g_{p_1 p_2 \cdots p_5}(t) = t \cdot g_{p_1 p_2 p_3 p_4}(t) + \sum_{r=1}^{4} (-1)^{r+1} \binom{4}{r} g_{\frac{p_1 p_2 p_3 p_4 p_5}{p_1 \cdots p_r}}(t)$$

$$= t(t + 7t^2 + 6t^3 + t^4) + \binom{4}{1} g_{p_2 \cdots p_5}(t) - \binom{4}{2} g_{p_3 p_4 p_5}(t) + \binom{4}{3} g_{p_4 p_5}(t) - \binom{4}{4} g_{p_5}(t)$$

$$= t(t + 7t^2 + 6t^3 + t^4) + 4(t + 7t^2 + 6t^3 + t^4) - 6(t + 3t^2 + t^3) + 4(t + t^2) - t$$

$$= t + 15t^2 + 25t^3 + 10t^4 + t^5. \quad \square$$

Example 8.8. Let $n = p_1^2 p_2^2$. Let q be a prime such that $(n, q) = 1$. Find $g_{nq}(t)$.

Solution. By Theorem 8.6, we have

$$g_{nq}(t) = t \cdot g_n(t) + \sum_{r=1}^{2}(-1)^{r+1}\binom{2}{r}g_{\frac{nq}{p_1\cdots p_r}}(t)$$

$$= t \cdot g_n(t) + \binom{2}{1}g_{\frac{nq}{p_1}}(t) - \binom{2}{2}g_{\frac{nq}{p_1 p_2}}(t)$$

$$= t \cdot g_n(t) + 2g_{p_1 p_2^2 q}(t) - g_{p_1 p_2 q}(t)$$

$$= t(t + 4t^2 + 3t^3 + t^4) + 2(t + 5t^2 + 4t^3 + t^4) - (t + 3t^2 + t^3)$$

$$= t + 8t^2 + 11t^3 + 5t^4 + t^5.$$

Therefore, $g_{nq}(t) = g_{p_1^2 p_2^2 p_3}(t) = t + 8t^2 + 11t^3 + 5t^4 + t^5$. $\square$

We can use Theorem 8.6 to prove an interesting identity for Stirling numbers of the second kind.

Theorem 8.9. *Let* $S(k,\alpha) = \begin{Bmatrix} k \\ \alpha \end{Bmatrix}$ *be a Stirling number of the second kind. Define*

$\Delta S(k,\alpha) = \begin{Bmatrix} k+1 \\ \alpha+1 \end{Bmatrix} - \begin{Bmatrix} k \\ \alpha \end{Bmatrix}$. *Then, for* $k > \alpha \geq 0$, *we have the following identity:*

$$\Delta S(k,\alpha) = \sum_{r=1}^{k-\alpha}(-1)^{r+1}\binom{k}{r}\begin{Bmatrix} k-r+1 \\ \alpha+1 \end{Bmatrix}.$$

Proof. For the special case of $m = 1$, Theorem 8.6 implies $n = p_1 p_2 \cdots p_k$, with $(n,q) = 1$, and

$g_{nq}(t) = t \cdot g_n(t) + \sum_{r=1}^{k}(-1)^{r+1}\binom{k}{r}g_{\frac{nq}{p_1 p_2 \cdots p_r}}(t)$. Since $[t^\beta]g_{p_1 p_2 \cdots p_k}(t) = \begin{Bmatrix} k \\ \beta \end{Bmatrix}$, we have

$[t^\beta]g_{nq}(t) = [t^\beta]\left(t \cdot g_n(t) + \sum_{r=1}^{k}(-1)^{r+1}\binom{k}{r}g_{\frac{nq}{p_1 p_2 \cdots p_r}}(t)\right)$. This implies that

$\begin{Bmatrix} k+1 \\ \beta \end{Bmatrix} = \begin{Bmatrix} k \\ \beta-1 \end{Bmatrix} + \sum_{r=1}^{k}(-1)^{r+1}\binom{k}{r}\begin{Bmatrix} k-r+1 \\ \beta \end{Bmatrix}$. Thus, for $k+1 > \beta \geq 1$, we have

$\begin{Bmatrix} k+1 \\ \beta \end{Bmatrix} - \begin{Bmatrix} k \\ \beta-1 \end{Bmatrix} = \sum_{r=1}^{k}(-1)^{r+1}\binom{k}{r}\begin{Bmatrix} k-r+1 \\ \beta \end{Bmatrix}$. Substituting $\beta = \alpha+1$, gives the result

$$\left\{{k+1\atop\alpha+1}\right\}-\left\{{k\atop\alpha}\right\}=\sum_{r=1}^{k}(-1)^{r+1}\binom{k}{r}\left\{{k-r+1\atop\alpha+1}\right\}.$$ Finally, since we must have $k-r+1\geq\alpha+1$, the summation only needs to be performed for $r=1$ to $r=k-\alpha$. This is because $\left\{{n\atop k}\right\}=0$ for $n\neq 0$ and $k>n$. □

Example 8.10. Let $k=4$ and $\alpha=2$. By definition, $\Delta S(4,2)=\left\{{5\atop 3}\right\}-\left\{{4\atop 2}\right\}=25-7=18$. Applying Theorem 8.9, we have

$$\Delta S(4,2)=\sum_{r=1}^{2}(-1)^{r+1}\binom{4}{r}\left\{{5-r\atop 3}\right\}$$

$$=(-1)^{2}\binom{4}{1}\left\{{4\atop 3}\right\}+(-1)^{3}\binom{4}{2}\left\{{3\atop 3}\right\}$$

$$=\binom{4}{1}\left\{{4\atop 3}\right\}-\binom{4}{2}\left\{{3\atop 3}\right\}$$

$$=4\cdot 6-6\cdot 1$$

$$=18.\ \square$$

Note that $\left\{{n\atop k}\right\}=0$ if $n,k\neq 0$ and $k>n$. Also, by definition, $\left\{{0\atop 0}\right\}=1$, and $\left\{{n\atop 0}\right\}=0$ for $n\neq 0$.

Corollary 8.11. For $k\geq m\geq 1$, $\left\{{k\atop m}\right\}=\dfrac{1}{m}\sum_{r=1}^{k-m+1}(-1)^{r+1}\binom{k}{r}\left\{{k-r+1\atop m}\right\}.$

Proof. Since the Stirling numbers of the second kind satisfy the recurrence relation

$\left\{{n\atop k}\right\}=k\left\{{n-1\atop k}\right\}+\left\{{n-1\atop k-1}\right\}$, or $\left\{{n\atop k}\right\}-\left\{{n-1\atop k-1}\right\}=k\left\{{n-1\atop k}\right\}$, we have $\Delta S(n-1,k-1)=k\left\{{n-1\atop k}\right\}$. Then, Theorem 8.9 implies

$$\Delta S(k,\alpha)=(\alpha+1)\left\{{k\atop\alpha+1}\right\}$$

$$=\sum_{r=1}^{k-\alpha}(-1)^{r+1}\binom{k}{r}\left\{{k+1-r\atop\alpha+1}\right\}.$$

If we let $m = \alpha + 1$, then we get

$$\Delta S(k, m-1) = m \begin{Bmatrix} k \\ m \end{Bmatrix}$$

$$= \sum_{r=1}^{k-m+1} (-1)^{r+1} \binom{k}{r} \begin{Bmatrix} k+1-r \\ m \end{Bmatrix}. \quad \square$$

Example 8.12. In Corollary 8.11, the quantity $\begin{Bmatrix} k \\ m \end{Bmatrix}$ is always an integer, because $\begin{Bmatrix} k \\ m \end{Bmatrix}$ is the number of non-empty disjoint m-sets formed from a non-empty k-set. Hence, m (≥ 1) must be a divisor of the quantity $\sum_{r=1}^{k-m+1} (-1)^{r+1} \binom{k}{r} \begin{Bmatrix} k+1-r \\ m \end{Bmatrix}$. This means that

$$\sum_{r=1}^{k-m+1} (-1)^{r+1} \binom{k}{r} \begin{Bmatrix} k+1-r \\ m \end{Bmatrix} \equiv 0 \pmod{m}.$$

For example, we have

$$\sum_{r=1}^{3} (-1)^{r+1} \binom{6}{r} \begin{Bmatrix} 7-r \\ 4 \end{Bmatrix} = (-1)^2 \binom{6}{1} \begin{Bmatrix} 6 \\ 4 \end{Bmatrix} + (-1)^3 \binom{6}{2} \begin{Bmatrix} 5 \\ 4 \end{Bmatrix} + (-1)^4 \binom{6}{3} \begin{Bmatrix} 4 \\ 4 \end{Bmatrix}$$

$$= \binom{6}{1} \begin{Bmatrix} 6 \\ 4 \end{Bmatrix} - \binom{6}{2} \begin{Bmatrix} 5 \\ 4 \end{Bmatrix} + \binom{6}{3} \begin{Bmatrix} 4 \\ 4 \end{Bmatrix}$$

$$= 6 \cdot 65 - 15 \cdot 10 + 20 \cdot 1$$

$$= 260.$$

And $260 \equiv 0 \pmod 4$, since 4 divides 260. $\square$

From Corollary 8.11, and as demonstrated in Example 8.12, we have the following beautiful result:

Corollary 8.13. *For $k \geq m \geq 1$, we have the following congruence, modulo m:*

$$\binom{k}{1}\begin{Bmatrix} k \\ m \end{Bmatrix} + \binom{k}{3}\begin{Bmatrix} k-2 \\ m \end{Bmatrix} + \binom{k}{5}\begin{Bmatrix} k-4 \\ m \end{Bmatrix} + \ldots \equiv \binom{k}{2}\begin{Bmatrix} k-1 \\ m \end{Bmatrix} + \binom{k}{4}\begin{Bmatrix} k-3 \\ m \end{Bmatrix} + \binom{k}{6}\begin{Bmatrix} k-5 \\ m \end{Bmatrix} + \ldots \pmod{m}.$$

Proof. In Corollary 8.11, the left-hand side of the equality is an integer. Therefore, on the right-hand side we must have $\sum_{r=1}^{k-m+1}(-1)^{r+1}\binom{k}{r}\left\{\dfrac{k+1-r}{m}\right\} \equiv 0 \pmod{m}$. We can now move the negative terms to the other side of the congruence. □

We can now use Theorem 8.6 to show how to add one ball (object) of a "like color" (or like kind) to a ball (object) specification of the form $n = p_1^{m-1} p_2$.

Theorem 8.14. *Let p_1 and p_2 be distinct primes. Then $g_{p_1^m p_2}(t) = g_{p_1^{m-1} p_2}(t) + t \cdot g_{p_1^m}(t)$.*

Proof. Let $n = p_1^m$. Let q be a prime other than p_1. Theorem 8.6 implies that

$$g_{p_1^m q}(t) = t \cdot g_{p_1^m}(t) + \sum_{r=1}^{1}(-1)^{r+1}\binom{1}{r} g_{\frac{p_1^m q}{p_1 \cdots p_r}}(t)$$

$$= t \cdot g_{p_1^m}(t) + g_{\frac{p_1^m q}{p_1}}(t)$$

$$= t \cdot g_{p_1^m}(t) + g_{p_1^{m-1} q}(t).$$

A simple change of variables ($q = p_2$) gives the desired result. □

Example 8.15. Given that $g_{p_1^4 p_2}(t) = t + 4t^2 + 4t^3 + 2t^4 + t^5$ and $g_{p_1^5}(t) = t + 2t^2 + 2t^3 + t^4 + t^5$, calculate $g_{p_1^5 p_2}(t)$. By Theorem 8.14 and using the polynomials in Appendix A, we have

$$g_{p_1^5 p_2}(t) = g_{p_1^4 p_2}(t) + t \cdot g_{p_1^5}(t)$$
$$= (t + 4t^2 + 4t^3 + 2t^4 + t^5) + t(t + 2t^2 + 2t^3 + t^4 + t^5)$$
$$= t + 5t^2 + 6t^3 + 4t^4 + 2t^5 + t^6. \quad □$$

Note that the polynomial $g_{p_1^5 p_2}(t)$ is the object distribution polynomial that counts, or enumerates, the number of ways to distribute five objects of one kind and one object of another kind into identical boxes with no box empty. The coefficient of t^k is the number of such distributions into precisely k identical boxes with no box empty.

Exercises

1. Use Lemma 8.4 to calculate $g_n(t)$ for $n = p_1 p_2 p_3$ where p_1, p_2, and p_3 are distinct primes.

2. Use Theorem 8.6 and Appendix A to calculate $g_n(t)$ for $n = p_1 p_2 p_3$ where p_1, p_2, and p_3 are distinct primes.

3. Use Corollary 8.13 to show that

$$\binom{8}{1}\left\{\begin{matrix}8\\3\end{matrix}\right\} + \binom{8}{3}\left\{\begin{matrix}6\\3\end{matrix}\right\} + \binom{8}{5}\left\{\begin{matrix}4\\3\end{matrix}\right\} \equiv \binom{8}{2}\left\{\begin{matrix}7\\3\end{matrix}\right\} + \binom{8}{4}\left\{\begin{matrix}5\\3\end{matrix}\right\} + \binom{8}{6}\left\{\begin{matrix}3\\3\end{matrix}\right\} \pmod{3}.$$

4. Use Theorem 8.14 to show that if p and q are distinct primes, then $g_{pq}(t) = g_q(t) + t \cdot g_p(t)$. Verify this result using the polynomials in Appendix A.

Chapter 9. Distributions in Boxes of Several Types

Consider the enumerator for distributing objects into identical boxes with no box empty:

$$G(x, t) = \prod_{d>1} \frac{1}{(1-tx^{\log d})} = \sum_{n=1}^{\infty} g_n(t) x^{\log n} \qquad (9.1)$$

Here, the boxes are all identical, or indistinguishable, and we can, for definiteness, think of them as white (w) boxes. Similarly, the same enumerator counts the distributions if all the boxes are, say, black (b). In (9.1), the variable t serves as a "counter" to count the number of boxes. If we want to emphasize that all of the boxes are colored white, then we can replace t by wt. Then w counts the white boxes and t counts the total number of boxes. So, for white boxes we have

$$G(x, wt) = \prod_{d>1} \frac{1}{(1-wtx^{\log d})} \qquad (9.2)$$

Similarly, for black boxes, we have

$$G(x, bt) = \prod_{d>1} \frac{1}{(1-btx^{\log d})} \qquad (9.3)$$

Now, what happens if we multiply $G(x, wt)$ and $G(x, bt)$? When we multiply and expand the product $G(x, wt) \cdot G(x, bt)$ we get many terms like the following:

$$(w^r t^r x^{\log n})(b^s t^s x^{\log m}) = w^r b^s t^{r+s} x^{\log(mn)} \qquad (9.4)$$

The term $w^r b^s t^{r+s} x^{\log(nm)}$ counts one way to distribute objects of specification n into r white boxes and objects of specification m into s black boxes. Equivalently, the term $w^r b^s t^{r+s} x^{\log(mn)}$ counts one way to distribute objects of specification $n \cdot m$ into $r+s$ boxes, of which r of the boxes are white and s of the boxes are black. Since the product $G(x, wt) \cdot G(x, bt)$ contains all possible combinations of terms like $w^r b^s t^{r+s} x^{\log(nm)}$, it follows that the product $G(x, wt) \cdot G(x, bt)$ enumerates all of the ways to distribute objects into two colors (e.g., white and black) of boxes. Thus, we have the following lemma.

Lemma 9.1. *The logarithmic generating function for counting the number of ways to distribute objects into two colors of boxes (e.g., white (w) and black (b)), with no box empty, is given by*

$$G(x, wt) \cdot G(x, bt) = \left(\prod_{d>1} \frac{1}{(1-wtx^{\log d})} \right) \left(\prod_{d>1} \frac{1}{(1-btx^{\log d})} \right)$$

$$= \sum_{n=1}^{\infty} \left(\sum_{d|n} g_d(wt) g_{\frac{n}{d}}(bt) \right) x^{\log n}$$

$$= \sum_{n=1}^{\infty} \left(\sum_{k} P_k(w, b) t^k \right) x^{\log n}.$$

The last part follows because whenever we multiply two logarithmic generating functions the terms generated are a convolution product (see Theorem 2.5). The polynomials $P_k(w, b)$ are homogeneous of degree k in the variables w and b.

Example 9.2. Suppose we have one red ball and one blue ball. Let the prime 2 represent the red ball, and let the prime 3 represent the blue ball. Then the specification number is $n = 2 \cdot 3 = 6$. Lemma 9.1 allows us to calculate the number of ways to distribute these two objects into two types of boxes, say white (w) and black (b) boxes. By Lemma 9.1, we have

$$\sum_{d|6} g_d(wt) g_{\frac{6}{d}}(bt) = g_1(wt)g_6(bt) + g_2(wt)g_3(bt) + g_3(wt)g_2(bt) + +g_6(wt)g_1(bt) =$$

$(w+b)t + (w^2 + 2wb + b^2)t^2$. (We have used the object distribution polynomials $g_n(t)$ that can be calculated from equation (9.1). Or, look up the polynomials in Appendix A.) So, $P_1(w, b) = w + b$ and $P_2(w, b) = w^2 + 2wb + b^2$. For two boxes, we look at the coefficient of t^2, which is the polynomial $P_2(w, b) = w^2 + 2wb + b^2$. Since the coefficient of w^2 is 1, there is one way to distribute objects of specification 6 into two white boxes with no box empty. Since the coefficient of wb is 2, there are two ways to distribute objects of specification 6 into one white box and one black box with no box empty. Finally, since the coefficient of b^2 is 1, there is one way to distribute objects of specification 6 into two black boxes with no box empty. Remember, the objects of specification 6 are one red ball and one blue ball.

Corollary 9.3. *If* $G(x, t) = \prod_{d>1} \frac{1}{(1-tx^{\log d})} = \sum_{n=1}^{\infty} g_n(t) x^{\log n}$ *is the logarithmic generating function for distributing objects into identical boxes with no box empty, then*

$$G^2(x, t) = \sum_{n=1}^{\infty} \left(\sum_{d|n} g_d(t) g_{\frac{n}{d}}(t) \right) x^{\log n}$$
$$= \sum_{n=1}^{\infty} \left(g_n(t) * g_n(t) \right) x^{\log n}.$$

Proof. Let $w = b = 1$ in Lemma 9.1. The convolution product is, by definition,

$$g_n(t) * g_n(t) = \sum_{d|n} g_d(t) g_{\frac{n}{d}}(t).$$

In other words, $G^2(x, t)$ is the logarithmic generating function for enumerating the number of ways to distribute objects into two kinds (or two "colors") of boxes with no box empty. And $g_n(t) * g_n(t)$ counts the number of ways to distribute objects of specification n into boxes of two colors, with no box empty.

Comment. In general, $g_n(t) * g_n(t)$ does *not* give us an object distribution polynomial for distributing objects of specification n into identical boxes with no box empty. There is no simple *closure law* for the convolution product of object distribution polynomials. However, as we have just seen, in Corollary 9.3, $g_n(t) * g_n(t)$ does give us an object distribution polynomial for distributing objects of specification n into *two kinds*, or colors, of boxes. So, convolution of object distribution polynomials jumps us up into a higher dimensional "box space."

Example 9.4. Let $n = 2 \cdot 3 = 6$, which is a specification number that encodes, say, one red ball and one blue ball. We have the following calculations (next page):

$$g_6(t) * g_6(t) = \sum_{d|n} g_d(t) g_{\frac{n}{d}}(t)$$
$$= g_1(t)g_6(t) + g_2(t)g_3(t) + g_3(t)g_2(t) + g_6(t)g_1(t)$$
$$= 1 \cdot (t + t^2) + t \cdot t + t \cdot t + (t + t^2) \cdot 1$$
$$= 2t + 4t^2$$

So, there are two ways to distribute one red ball and one blue ball into one box (the coefficient of t is 2) which can be either a white box or a black box:

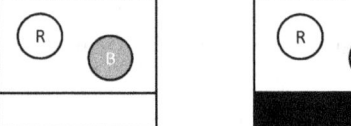

Also, since the coefficient of t^2 is 4, there are four ways to distribute one red ball and one blue ball into two boxes which can be either white or black in any combination:

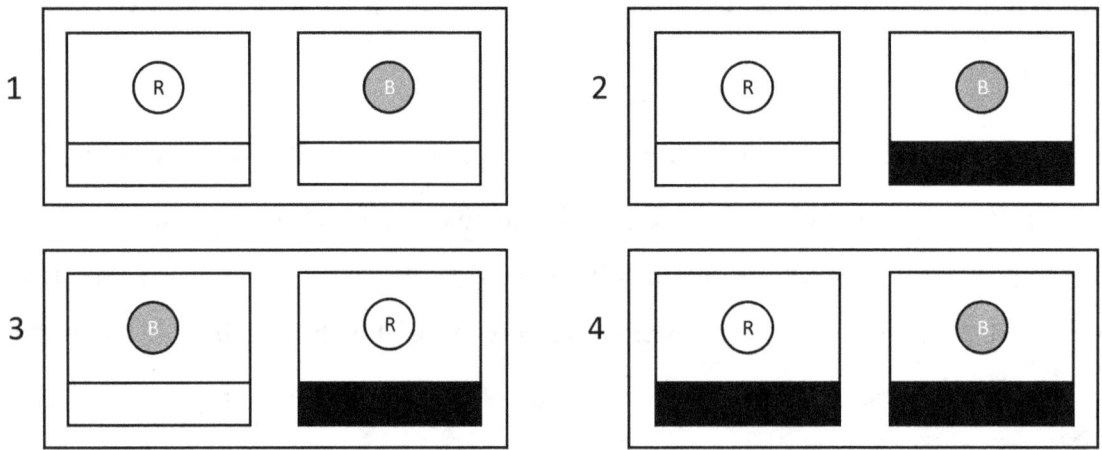

Corollary 9.3 can easily be generalized as follows.

Theorem 9.5. *If* $G(x, t) = \prod_{d>1} \frac{1}{(1-tx^{\log d})} = \sum_{n=1}^{\infty} g_n(t) x^{\log n}$, *then the following generating function enumerates the number of ways to distribute objects of specification n into boxes of r different types (or "colors"), in all possible combinations, with no box empty:*

$$G^r(x, t) = \left(\prod_{d>1} \frac{1}{(1-tx^{\log d})} \right)^r \qquad (9.5)$$
$$= \sum_{n=1}^{\infty} \left(\sum_{n_1 n_2 \ldots n_r = n} g_{n_1}(t) g_{n_2}(t) \cdots g_{n_r}(t) \right) x^{\log n}.$$

Proof. Use induction on Corollary 9.3, remembering that multiplying logarithmic generating functions results in taking convolution products of the coefficients of $x^{\log n}$. □

Example 9.6. In how many ways can we distribute two red balls and one blue ball into any combination of boxes that come in two colors, say white (w) and black (b)?

Solution. Let $n = 2^2 \cdot 3 = 12$ where the prime 2 represents a red ball and the prime 3 represents a blue ball. So, $\{r, r, b\} = \{2, 2, 3\}$ and $n = 2 \cdot 2 \cdot 3 = 12$ is the specification number for the objects (colored balls). For boxes of two colors, we let $r = 2$. Then, by Theorem 9.5, the coefficient of $x^{\log 12}$ is

$$\sum_{n_1 n_2 = 12} g_{n_1}(wt) g_{n_2}(bt) =$$
$$= g_1(wt)g_{12}(bt) + g_2(wt)g_6(bt) + g_3(wt)g_4(bt) + g_4(wt)g_3(bt) + g_6(wt)g_2(bt) + g_{12}(wt)g_1(bt)$$
$$= (w+b)t + (2w^2 + 4wb + 2b^2)t^2 + (w^3 + 2w^2 b + 2wb^2 + b^3)t^3.$$

(The polynomials $g_n(t)$ are obtained from equation (9.1).) We can now read off the number of distributions for a given number of boxes and types. For example, the coefficient of t^2, for two boxes, is $2w^2 + 4wb + 2b^2$. If we have one white box and one black box, then we note that the coefficient of wb is 4. In fact, the four distributions into one white box and one black box are as follows:

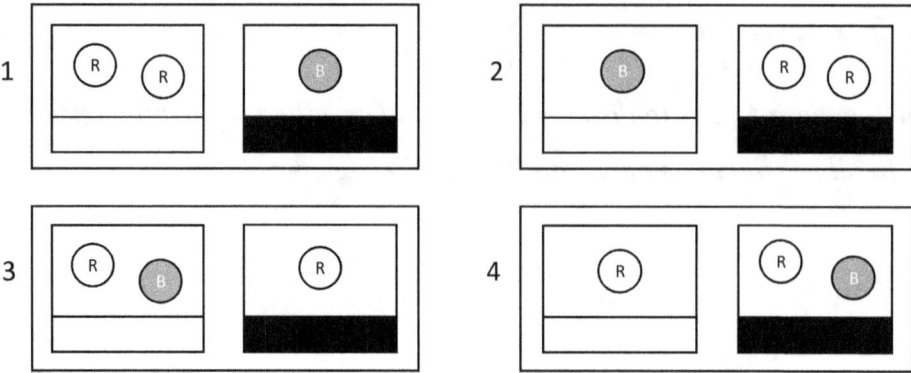

Theorem 9.7. *The logarithmic generating function for enumerating the number of distributions of objects, of specification n, into boxes of r colors in all possible combinations, with no box empty, is given by*

$$\prod_{d>1}\left(\sum_{k=0}^{\infty}\binom{k+r-1}{k}t^k x^{k\log d}\right) = \sum_{n=1}^{\infty}\left(\sum_{n_1 n_2 \ldots n_r = n} g_{n_1}(t)g_{n_2}(t)\ldots g_{n_r}(t)\right)x^{\log n}. \tag{9.6}$$

Proof. Let $u = tx^{\log d}$. Then by the General Binomial Theorem, we have $\dfrac{1}{(1-u)^r} = \sum_{k=0}^{\infty}\binom{k+r-1}{k}u^k$.

Substituting this into Theorem 9.5 gives the result in Theorem 9.7. □

Theorem 9.7 is a useful equivalent representation of Theorem 9.5. The significance of Theorem 9.7 is that it tells us how to "build" solutions to object distribution problems. By identifying the *ordered* factorizations of *n*, we can build a solution to the problem of enumerating the number of ways to distribute objects of specification *n* into boxes from an unlimited supply of *r* colors, taken over all possible combinations of boxes, with no box empty.

In Theorem 9.7, the coefficient of each term of the form $x^{k\log d}$, or $x^{\log d^k}$, is $\binom{k+r-1}{k}t^k$. When we expand the left-hand side of (9.6), the coefficient of $x^{\log n}$ is obtained from all *ordered* factorizations of *n* having the form $d_1^{k_1} d_2^{k_2} \ldots d_m^{k_m}$ where $d_i \neq d_j$, for $i \neq j$, and $1 < d_1 < d_2 < \ldots < d_m$ and $d_i^{k_i}$ divides *n*. Suppose that we have objects of specification *n* and we want to calculate the number of ways to distribute the objects into boxes of *r* colors, with color repetition allowed. The solution will be a polynomial $g_n(r, t)$ where the coefficient of t^k is the number of ways to place the objects into k boxes whose colors are chosen from the set of r colors with repetition allowed. We can use Theorem 9.7 to build the polynomial $g_n(r, t)$. For each valid factorization of the form $d_1^{k_1} d_2^{k_2}...d_m^{k_m}$, we add a term of the form $\binom{k_1+r-1}{k_1}\binom{k_2+r-1}{k_2}...\binom{k_m+r-1}{k_m} t^{k_1+k_2+...+k_m}$. Then we sum all the terms to get $g_n(r, t)$. This method is demonstrated in Example 9.13.

While these observations may seem mysterious, they are actually quite simple. If we expand the left-hand side of (9.6), using actual numbers for $d = 2, 3, 4, ...$, we will observe the aforementioned properties. A couple examples suffice to illustrate Theorem 9.7.

Example 9.8. Suppose we have four identical objects. Using the prime number 2 to represent one of the objects, we can represent the four identical objects by the specification number $n = 2^4 = 16$. The divisors of 16 are 1, 2, 4, 8, and 16. We next find the factorizations of 16 having the form $d_1^{k_1} d_2^{k_2}...d_m^{k_m}$ where $d_i \neq d_j$ for $i \neq j$, and $1 < d_1 < d_2 < ... < d_m$ and $d_i^{k_i}$ divides 16. One way to do this is to multiply the product $(1+2^1+2^2+2^3+2^4)(1+4^1+4^2)(1+8^1)(1+16^1)$ and look for the factorizations of 16 with the desired properties. There are five such factorizations: 16^1, $2^1 \cdot 8^1$, $2^2 \cdot 4^1$, 2^4, and 4^2. Note that we cannot have a factorization of 16 like $2^2 \cdot 2^2$, since expansion of the left-hand side of (9.6) does *not* give a product of $t^2 x^{\log 2^2}$ with $t^2 x^{\log 2^2}$. Next, for each valid factorization, like $d_1^{k_1} d_2^{k_2}...d_m^{k_m}$, we form the term

$$\binom{k_1+r-1}{k_1} t^{k_1} \cdot \binom{k_2+r-1}{k_2} t^{k_2} ... \binom{k_m+r-1}{k_m} t^{k_m}$$

and we sum all the terms. In this example, 16^1, $2^1 \cdot 8^1$, $2^2 \cdot 4^1$, 2^4, and 4^2 give the polynomial $g_n(r, t) = g_{16}(r, t)$:

$$g_{16}(r, t) = \binom{r}{1} t^1 + \binom{r}{1}\binom{r}{1} t^1 + \binom{r+1}{2} t^2 \binom{r}{1} t^1 + \binom{r+3}{4} t^4 + \binom{r+1}{2} t^2$$

$$= rt + (\tfrac{1}{2}r + \tfrac{3}{2}r^2)t^2 + (\tfrac{1}{2}r^2 + \tfrac{1}{2}r^3)t^3 + (\tfrac{1}{4}r + \tfrac{11}{24}r^2 + \tfrac{1}{4}r^3 + \tfrac{1}{24}r^4)t^4.$$

The polynomial $g_{16}(r, t)$ tells us the number of ways to distribute four identical objects into boxes of r colors, for all possible combinations of the colors, with no box empty. To see how to use the polynomial $g_{16}(r, t)$, suppose we have two boxes. For two boxes, the coefficient of t^2 is $\frac{1}{2}r + \frac{3}{2}r^2$. For $r = 2$ colors for boxes, we have $\frac{1}{2}r + \frac{3}{2}r^2 = \frac{1}{2} \cdot 2 + \frac{3}{2} \cdot 2^2 = 7$. In fact, the 7 distributions are the following:

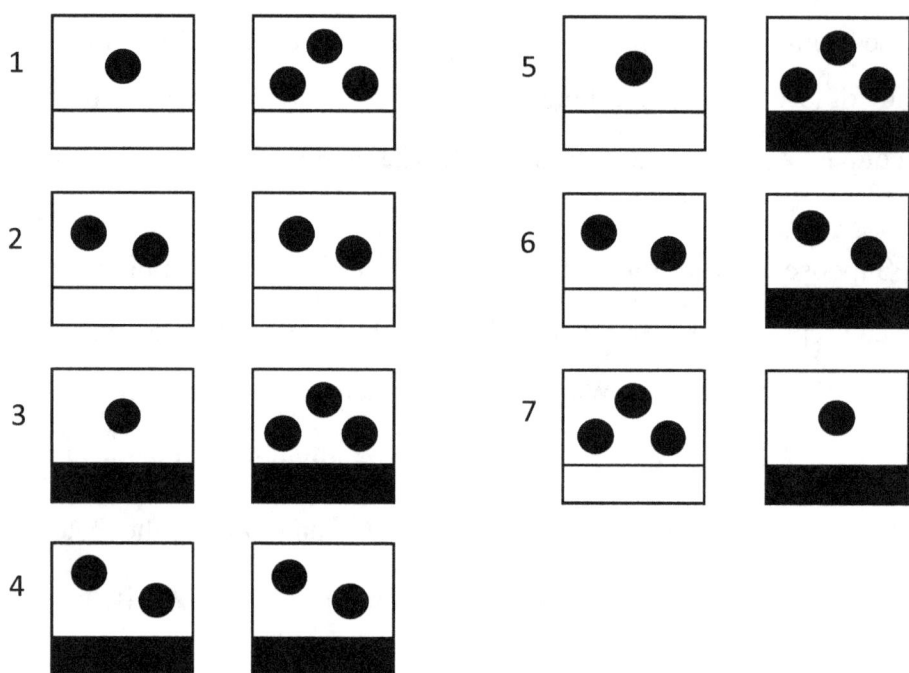

Example 9.8 provides some interesting insights into the distribution problem for the case where all the objects are identical. For $n = 16$, the divisors of the form $d_1^{k_1} d_2^{k_2} \ldots d_m^{k_m}$ where $d_i \neq d_j$ for $i \neq j$, and $1 < d_1 < d_2 < \ldots < d_m$ are 16^1, $2^1 \cdot 8^1$, $2^2 \cdot 4^1$, 2^4, and 4^2. We can write these factorizations as 16, $2 \cdot 8$, $2 \cdot 2 \cdot 4$, $2 \cdot 2 \cdot 2 \cdot 2$, and $4 \cdot 4$. And these factorizations can be written as (2^4), $(2^1)(2^3)$, $(2^1)(2^1)(2^2)$, $(2^1)(2^1)(2^1)(2^1)$, and $(2^2)(2^2)$. We now observe that the sum of exponents in each of these factorizations is a partition of the number 4: 4, $1+3$, $1+1+2$, $1+1+1+1$, and $2+2$. In fact, these are the five integer partitions of the number 4. So, $p(4) = 5$. These five partitions of 4 have the types $[4^1]$, $[1^1 3^1]$, $[1^2 2^1]$, $[1^4]$, and $[2^2]$. This observation provides another way to build the polynomial $g_n(r, t)$ for identical objects by using the exponents in each of the partition types $[1^{k_1} 2^{k_2} \ldots m^{k_m}]$.

Suppose we have *m identical* objects and we want to construct the distribution polynomial $g(r, t)$ that counts the number of ways to distribute the m objects into boxes of r colors, with no box empty. For each integer partition of m having the type $[1^{k_1} 2^{k_2} ... m^{k_m}]$, we add a term to $g(r, t)$ of the form

$$\binom{k_1 + r - 1}{k_1}\binom{k_2 + r - 1}{k_2}...\binom{k_m + r - 1}{k_m} t^{k_1 + k_2 + ... + k_m}. \tag{9.7}$$

Then $g(r, t)$ for m identical objects is the sum of all terms of the form (9.7) taken over all integer partitions of m. We can state this as the following theorem.

Theorem 9.9. *Let $g(m, r, t)$ denote the object distribution polynomial that counts the number of ways to distribute m identical objects into boxes of r colors, using all possible combinations of colors, with color repetition allowed, and no box empty. Then*

$$g(m, r, t) = \sum_{[1^{k_1} 2^{k_2} ... m^{k_m}]} \binom{k_1 + r - 1}{k_1}\binom{k_2 + r - 1}{k_2}...\binom{k_m + r - 1}{k_m} t^{k_1 + k_2 + ... + k_m}, \tag{9.8}$$

where the sum is taken over all positive integer partitions $[1^{k_1} 2^{k_2} ... m^{k_m}]$ of m.

Corollary 9.10. *The number of ways to color m identical objects using r colors, with repetition of colors allowed, is $\binom{m + r - 1}{m}$.*

Proof 1. As a corollary of Theorem 9.9, note that if we place m identical objects into m boxes, then each box will contain precisely one object. If the color of the box is transferred to the object, then we are effectively "painting," or coloring, the object with that color. The number of ways that we can thus color the m objects is given by the coefficient of t^m in Theorem 9.9. The term that generates the coefficient of t^m corresponds to the partition $[1^m 2^0 ... m^0]$, so the number of colorings is

$$\binom{m + r - 1}{m}\binom{0 + r - 1}{0}...\binom{0 + r - 1}{0} = \binom{m + r - 1}{m}. \square$$

Proof 2. The number of ways to choose m colors from a set of r colors, with repetition allowed, is
$$C_R(r, m) = \binom{r+m-1}{m}. \quad \square$$

Example 9.11. Suppose we have $m = 5$ identical objects. There are seven partitions of 5, namely $5, 1+4, 2+3, 1+1+3, 1+2+2, 1+1+1+2,$ and $1+1+1+1+1$. These seven partitions have types $[5^1], [1^1 4^1], [2^1 3^1], [1^2 3^1], [1^1 2^2], [1^3 2^1]$, and $[1^5]$. By Theorem 9.9, we have

$$g(5, r, t) = \binom{r}{1}t + \binom{r}{1}\binom{r}{1}t^2 + \binom{r}{1}\binom{r}{1}t^2 + \binom{r+1}{2}\binom{r}{1}t^3 + \binom{r}{1}\binom{r+1}{2}t^3 + \binom{r+2}{3}\binom{r}{1}t^4 + \binom{r+4}{5}t^5.$$

We can obviously expand this expression into a polynomial in the variables r and t. If, for example, we have only one color of box, then $r = 1$, and $g(5, 1, t) = t + 2t^2 + 2t^3 + t^4 + t^5$. Note that $g(5, 1, 1) = p(5) = 7$, and, in general, $g(m, 1, 1) = p(m)$, the number of partitions of m.

Example 9.12. Suppose we have two red balls and two blue balls. We can encode these objects by the specification number $n = 2^2 \cdot 3^2 = 36$. Using Theorem 9.7, or, equivalently, identifying the factorizations of $n = 36$ having the form $d_1^{k_1} d_2^{k_2} \ldots d_m^{k_m}$ where $d_i \neq d_j$ for $i \neq j$, and $1 < d_1 < d_2 < \ldots < d_m$ we get the factorizations $36, 6^2, 4 \cdot 9, 2^2 \cdot 9, 3^2 \cdot 4, 2^2 \cdot 3^2, 2 \cdot 18, 3 \cdot 12$, and $2 \cdot 3 \cdot 6$. Then

$$g_{36}(r, t) =$$
$$= \binom{r}{1}t + \binom{r+1}{2}t^2 + \binom{r}{1}\binom{r}{1}t^2 + \binom{r+1}{2}\binom{r}{1}t^3 + \binom{r+1}{2}\binom{r}{1}t^3 + \binom{r+1}{2}\binom{r+1}{2}t^4$$
$$+ \binom{r}{1}\binom{r}{1}t^2 + \binom{r}{1}\binom{r}{1}t^2 + \binom{r}{1}\binom{r}{1}\binom{r}{1}t^3$$
$$= rt + (\tfrac{1}{2}r + \tfrac{7}{2}r^2)t^2 + (r^2 + 2r^3)t^3 + (\tfrac{1}{4}r^2 + \tfrac{1}{2}r^3 + \tfrac{1}{4}r^4)t^4.$$

So, if we want to distribute two red balls and two blue balls into, say, two boxes, we look at the coefficient of t^2, which is $\tfrac{1}{2}r + \tfrac{7}{2}r^2$. Here, r is the number of colors that the boxes can be. If we have boxes of two colors, say black and white, then $r = 2$ and $g_{36}(2, 2) = 15$. This means that the two boxes are either black-and-black, black-and-white, or white-and-white (the order of the boxes does not matter). And there are 15 ways to distribute the objects into these combinations of boxes:

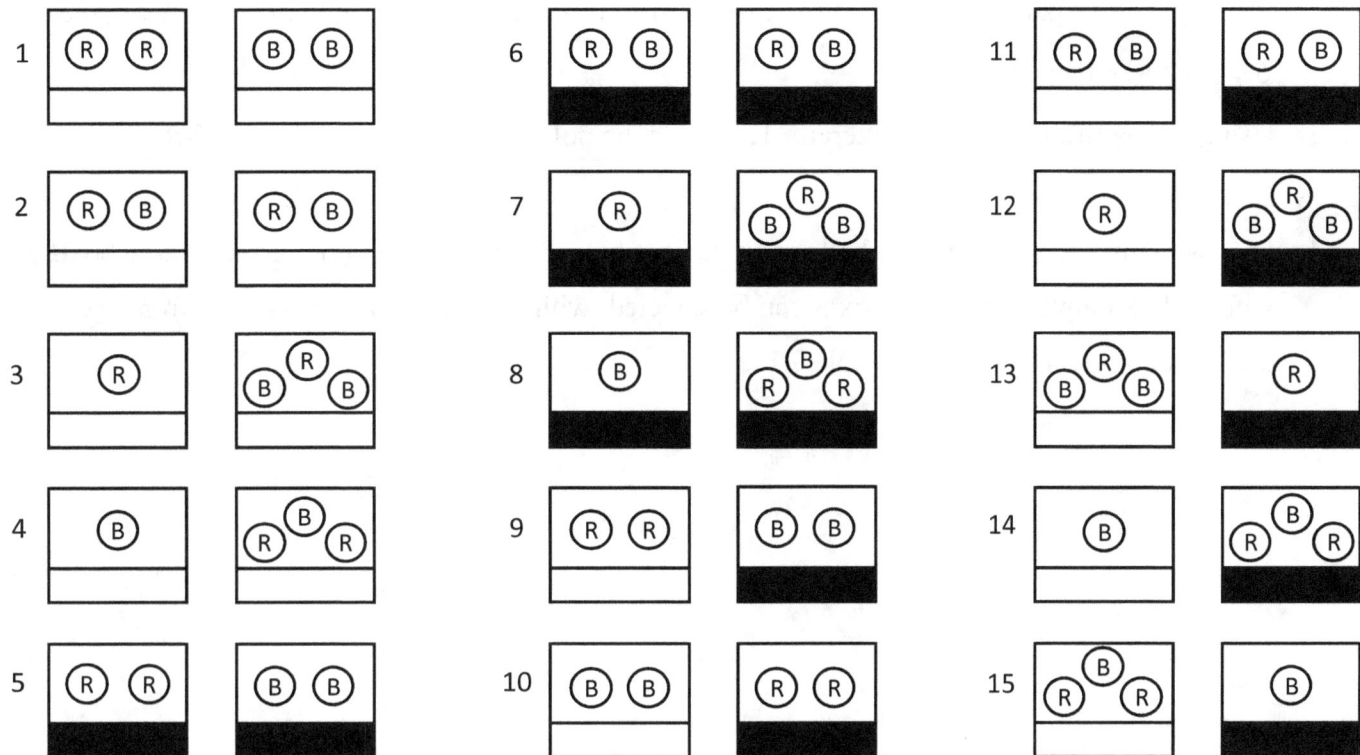

Exercises

1. Calculate $P(t) = g_8(t) * g_8(t)$. What is the combinatorial meaning of the polynomial $P(t)$?

2. What are the 11 partitions of 6? Write these partitions using the "type" notation. For example, $6 = 2+2+1+1 = [1^2 2^2]$.

3. Using the partitions of 6 from exercise 1, what is the polynomial $g(6, r, t)$? Hint: See Example 9.11.

4. In how many ways can two red balls and two blue balls be distributed into precisely three boxes, with no box empty, where the boxes can be selected, with color repetitions allowed, from six different colors. Hint: See Example 9.12.

Chapter 10. Distributions in Distinct Boxes

Suppose that we have a set of objects, such as colored balls, and we want to count the number of ways to distribute these balls into *distinct*, or labeled, boxes with no box empty. (The case of empty boxes is easily handled, so we focus on no empty boxes allowed.) For example, we might have three red balls, two blue balls, and five white balls, and we want to count the number of ways that we can distribute this set of colored balls into, say, four distinct boxes. Conventional mathematical theory provides us with an answer to this problem. For reference, I will present the standard textbook solution so that you can compare it with my formula, derived using the number theoretic paradigm, presented in Theorem 10.2. The following standard theorem is from Charalambides [2].

Reference Theorem A. *The number of distributions of n balls of specification* $[m_1, m_2, ..., m_n]$, *with* $m_i \geq 0$, $i = 1, 2, ..., n$ *and* $m_1 + 2m_2 + ... + nm_n = n$, *into k distinct boxes so that r of the boxes are occupied is given by*

$$R(m_1, m_2, ..., m_n; k, r) = \binom{k}{r} \sum_{j=0}^{r} (-1)^j \binom{r}{j} U(m_1, m_2, ..., m_n; r-j)$$

where $U(m_1, m_2, ..., m_n; k) = \binom{k}{1}^{m_1} \binom{k+1}{2}^{m_2} \cdots \binom{k+n-1}{n}^{m_n}$.

Theorem 10.2, which calculates the same numerical answers, appears to be much simpler because we will use the number-theoretic paradigm where the set of objects is encoded into a single specification number n.

The enumeration logarithmic generating function for counting the number of ways to distribute objects of specification n into identical boxes with no box empty is given by:

$$G(x, t) = \sum_{n=1}^{\infty} g_n(t) x^{\log n} = \prod_{d>1} \frac{1}{(1 - tx^{\log d})} \qquad (10.1)$$

Let $N(n, m)$ denote the number of ways to distribute objects of specification n into m distinct boxes with no box empty. Before deriving a general formula for $N(n, m)$, we will first need the following lemma. The lemma is rather tedious to prove, but it is required for my derivation of a formula for $N(n, m)$.

Lemma 10.1. *Let the specification number for a set of objects be n. The number of ways to distribute the objects into precisely m distinct boxes, with no box empty, is given by*

$$N(n, m) = \sum_{\substack{d_{m-1}|n \\ d_{m-1} \neq 1, n}} \sum_{\substack{d_{m-2}|d_{m-1} \\ d_{m-2} \neq 1, d_{m-1}}} \cdots \sum_{\substack{d_1|d_2 \\ d_1 \neq 1, d_2}} 1 \qquad (10.2)$$

This is not a pretty formula, and it would be unwieldy if we had to use it to calculate $N(n, m)$. Fortunately, the only reason I state this lemma is so that we can derive a much nicer formula for $N(n, m)$.

Proof. For m distinct boxes and objects of specification n, we have

$$\sum_{n=1}^{\infty} g_n(t, \alpha_1, \alpha_2, \ldots, \alpha_m) x^{\log n} = \prod_{d=2}^{\infty} \frac{1}{(1-t\alpha_1 x^{\log d})(1-t\alpha_2 x^{\log d})\ldots(1-t\alpha_m x^{\log d})}$$

$$= \prod_{d=2}^{\infty} \frac{1}{(1-t\alpha_1 x^{\log d})} \prod_{d=2}^{\infty} \frac{1}{(1-t\alpha_2 x^{\log d})} \cdots \prod_{d=2}^{\infty} \frac{1}{(1-t\alpha_m x^{\log d})}$$

$$= \left(\sum_{n=1}^{\infty} g_n(t\alpha_1) x^{\log n}\right)\left(\sum_{n=1}^{\infty} g_n(t\alpha_2) x^{\log n}\right)\cdots\left(\sum_{n=1}^{\infty} g_n(t\alpha_m) x^{\log n}\right)$$

$$= \sum_{n=1}^{\infty} (g_n(t\alpha_1) * g_n(t\alpha_2) * \ldots * g_n(t\alpha_m)) x^{\log n}$$

This is a generating function, so by comparing the coefficients of $x^{\log n}$ on both sides of the equation, we see that

$$g_n(t, \alpha_1, \alpha_2, ..., \alpha_m) = g_n(t\alpha_1) * g_n(t\alpha_2) * ... * g_n(t\alpha_m)$$

$$= \sum_{d_1 d_2 ... d_m = n} g_{d_1}(t\alpha_1) g_{d_2}(t\alpha_2) ... g_{d_m}(t\alpha_m) \quad (10.3)$$

$$= \sum_{\substack{d_{m-1}|n \\ d_{m-1} \neq 1, n}} \left(\sum_{\substack{d_{m-2}|d_{m-1} \\ d_{m-2} \neq 1, d_{m-1}}} ... \left(\sum_{\substack{d_1 | d_2 \\ d_1 \neq 1, d_2}} g_{d_1}(t\alpha_1) g_{\frac{d_2}{d_1}}(t\alpha_2) \right) ... g_{\frac{d_{m-1}}{d_{m-2}}}(t\alpha_{m-1}) \right) g_{\frac{n}{d_{m-1}}}(t\alpha_m)$$

If we want to count the number of ways to distribute objects of specification n into m distinct boxes, then we expand the right-hand side of (10.3) and look at the coefficient of $\alpha_1 \alpha_2 ... \alpha_m t^m$. When we do this, we will get a term of the form $\alpha_1 \alpha_2 ... \alpha_m t^m$ for each divisor d_j where d_j is neither 1 nor d_{j+1}. This is because $g_1(t\alpha_k) = 1$ always, and $g_{\frac{d_{j+1}}{d_{j+1}}}(t\alpha_k) = g_1(t\alpha_k) = 1$. In these cases, the product terms would not have a $t\alpha_k$ factor. Thus, the coefficient of $\alpha_1 \alpha_2 ... \alpha_m t^m$ in the expansion of the right-hand side of (10.3) is $\sum_{\substack{d_{m-1}|n \\ d_{m-1} \neq 1, n}} \sum_{\substack{d_{m-2}|d_{m-1} \\ d_{m-2} \neq 1, d_{m-1}}} ... \sum_{\substack{d_1|d_2 \\ d_1 \neq 1, d_2}} 1$. And, as we mentioned previously, the coefficient of $\alpha_1 \alpha_2 ... \alpha_m t^m$ in the expansion of (10.3) is $N(n, m)$, the number of ways to distribute objects of specification n into precisely m distinct boxes. □

Theorem 10.2. *Let n be the specification number for a set of objects where the total number of objects is at least m. The number of ways to distribute the objects into precisely m distinct boxes, with no box empty, is given by*

$$N(n, m) = \sum_{k=0}^{m-1} (-1)^k \binom{m}{k} \tau_{m-k-1}(n), \quad (10.4)$$

where $\tau_0(n) = 1$ and $\tau_{r+1}(n) = \sum_{d|n} \tau_r(d)$.

Compare equation (10.4) with the equivalent formula given in Reference Theorem A (different notations are used, so don't confuse the parameters n and m in the two formulas).

Proof. The author discovered (10.4) by doing actual calculations using (10.2) in Lemma 10.1. Using Lemma 10.1, we discover the first few simple cases of $N(n, m)$:

- $N(n, 1) = 1 = \tau_0(n)$

- $N(n, 2) = \sum_{\substack{d|n \\ d \neq 1, n}} 1 = \tau(n) - 2 = \tau_1(n) - 2\tau_0(n)$

- $N(n, 3) = \sum_{\substack{d_2|n \\ d_2 \neq 1, n}} \sum_{\substack{d_1|d_2 \\ d_1 \neq 1, d_2}} 1 = \sum_{d_2|n} \tau(d_2) - 3\tau(n) + 3 = \tau_2(n) - 3\tau_1(n) + 3\tau_0(n)$

From these three simple examples, we conjecture that $N(n, m) = \sum_{k=0}^{m-1}(-1)^k \binom{m}{k} \tau_{m-k-1}(n)$, and we can prove this using mathematical induction. We have already observed, above, that equation (10.4) is true for the initial cases $m = 1, 2, 3$. This establishes the induction basis. Next, we must show that if $N(n, m)$ is given by (10.4), then $N(n, m+1)$ is also given by (10.4). That will prove the theorem. So, assume that (10.4) is true for some positive integer m. By Lemma 10.1, we have

$$N(n, m+1) = \sum_{\substack{d_m|n \\ d_m \neq 1, n}} \left(\sum_{\substack{d_{m-1}|d_m \\ d_{m-1} \neq 1, d_m}} \sum_{\substack{d_{m-2}|d_{m-1} \\ d_{m-2} \neq 1, d_{m-1}}} \cdots \sum_{\substack{d_1|d_2 \\ d_1 \neq 1, d_2}} 1 \right)$$

$$= \sum_{\substack{d_m|n \\ d_m \neq 1, n}} N(d_m, m) = \sum_{\substack{d|n \\ d \neq 1, n}} N(d, m)$$

$$= \sum_{\substack{d|n \\ d \neq 1, n}} \left(\sum_{k=0}^{m-1} (-1)^k \binom{m}{k} \tau_{m-k-1}(d) \right)$$

$$= \sum_{d|n} \left(\sum_{k=0}^{m-1} (-1)^k \binom{m}{k} \tau_{m-k-1}(d) \right) - \sum_{k=0}^{m-1} (-1)^k \binom{m}{k} \tau_{m-k-1}(1) - \sum_{k=0}^{m-1} (-1)^k \binom{m}{k} \tau_{m-k-1}(n)$$

$$= \sum_{k=0}^{m-1} (-1)^k \binom{m}{k} \sum_{d|n} \tau_{m-k-1}(d) - \sum_{k=0}^{m-1} (-1)^k \binom{m}{k} - \sum_{k=0}^{m-1} (-1)^k \binom{m}{k} \tau_{m-k-1}(n)$$

$$= \sum_{k=0}^{m-1} (-1)^k \binom{m}{k} \tau_{m-k}(n) - \sum_{k=0}^{m-1} (-1)^k \binom{m}{k} - \sum_{k=0}^{m-1} (-1)^k \binom{m}{k} \tau_{m-k-1}(n)$$

$$= \sum_{k=0}^{m} (-1)^k \binom{m}{k} \tau_{m-k}(n) - (-1)^m \binom{m}{m} \tau_0(n) - \sum_{k=0}^{m-1} (-1)^k \binom{m}{k} - \sum_{k=0}^{m-1} (-1)^k \binom{m}{k} \tau_{m-k-1}(n)$$

$$= \sum_{k=0}^{m} (-1)^k \tau_{m-k}(n) \left(\binom{m+1}{k} - \binom{m}{k-1} \right) - (-1)^m - \sum_{k=0}^{m-1} (-1)^k \binom{m}{k} - \sum_{k=0}^{m-1} (-1)^k \binom{m}{k} \tau_{m-k-1}(n)$$

$$= \sum_{k=0}^{m} (-1)^k \binom{m+1}{k} \tau_{m-k}(n) - \sum_{k=0}^{m} (-1)^k \binom{m}{k-1} \tau_{m-k}(n) - (-1)^m - \sum_{k=0}^{m-1} (-1)^k \binom{m}{k}$$

$$- \sum_{k=0}^{m-1} (-1)^k \binom{m}{k} \tau_{m-k-1}(n)$$

$$= \sum_{k=0}^{m}(-1)^k\binom{m+1}{k}\tau_{m-k}(n) + \sum_{r=0}^{m-1}(-1)^r\binom{m}{r}\tau_{m-r-1}(n) - (-1)^m - \sum_{k=0}^{m-1}(-1)^k\binom{m}{k} - \sum_{k=0}^{m-1}(-1)^k\binom{m}{k}\tau_{m-k-1}(n)$$

$$= \sum_{k=0}^{m}(-1)^k\binom{m+1}{k}\tau_{m-k}(n) - (-1)^m - \sum_{k=0}^{m-1}(-1)^k\binom{m}{k}$$

$$= \sum_{k=0}^{m}(-1)^k\binom{m+1}{k}\tau_{m-k}(n), \text{ since } -(-1)^m - \sum_{k=0}^{m-1}(-1)^k\binom{m}{k} = 0 \text{ for } m \geq 1. \square$$

The following corollary to Theorem 10.2 illustrates the interplay between combinatorics and number theory. Theorem 10.2 has both a combinatorial and a number-theoretic interpretation.

Corollary 10.3. *If the number of primes in the prime factorization of n is less than m, then*

$$\sum_{k=0}^{m-1}(-1)^k\binom{m}{k}\tau_{m-k-1}(n) = 0. \tag{10.5}$$

Proof. The number of primes in the factorization of n is just the total number of objects that we are distributing into m distinct boxes with no box empty. If there are fewer objects than boxes, then we cannot possibly distribute the objects into the boxes with no box empty. $\square$

Theorem 10.2 introduced the generalized divisor function $\tau_k(n)$ which is defined recursively by $\tau_0(n) = 1$ and $\tau_{r+1}(n) = \sum_{d|n}\tau_r(d)$. This function has some beautiful mathematical properties that we will discuss later in this chapter. Using the recursive definition, we can easily build a reference table of $\tau_k(n)$ to facilitate calculations using Theorem 10.2. See Table 10.1 on page 174. Later, we will derive some properties of $\tau_k(n)$ that allow us to calculate its values directly, without having to use the defining recurrence relations or a reference table.

Example 10.4. Here is a simple example of how to use Theorem 10.2. Suppose we have two red balls and two blue balls. In how many ways can we distribute these colored balls into precisely three *distinct*, or labeled, boxes with no box empty? First, we build a specification number for the set of objects. Pick any two distinct prime numbers, say 2 and 3. Let the prime 2 represent a red ball and let 3 represent a blue ball. Our multiset of objects is then S = {red, red, blue, blue} = {R, R, B, B} = {2, 2, 3, 3}. The specification number of the set of objects is $n = 2 \cdot 2 \cdot 3 \cdot 3 = 36$. So, by Theorem 10.2, and using $\tau_k(n)$ values from Table 10.1, page 174, we have

$$N(36, 3) = \sum_{k=0}^{2} (-1)^k \binom{3}{k} \tau_{2-k}(36)$$

$$= (-1)^0 \binom{3}{0} \tau_2(36) + (-1)^1 \binom{3}{1} \tau_1(36) + (-1)^2 \binom{3}{2} \tau_0(36)$$

$$= \tau_2(36) - 3\tau_1(36) + 3\tau_0(36)$$

$$= 36 - (3)(9) + (3)(1)$$

$$= 12$$

So, the mathematics tells us that there are 12 ways to distribute the objects into the boxes according to the given specifications. The 12 ways are shown in Figure 10.1, below.

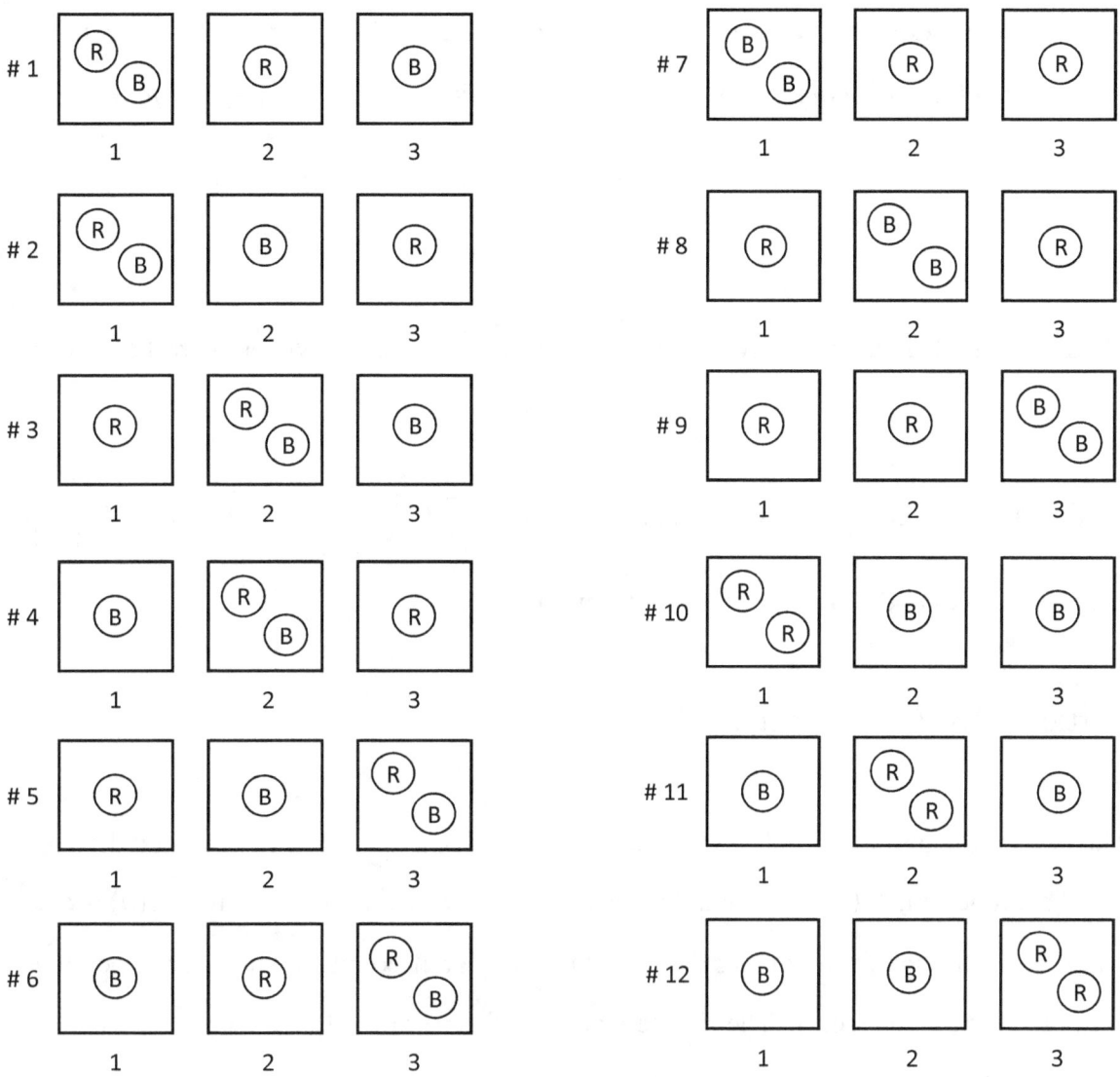

Figure 10.1. The 12 object distributions calculated in Example 10.4.

Using the technique of "inversion," we can also use Theorem 10.2 to calculate $\tau_k(n)$ in terms of the number of object distributions $N(n, m)$. The following lemma will be useful later when we further study the function $\tau_k(n)$ and derive some of its properties.

Lemma 10.5 (Inversion of Theorem 10.2). *The generalized divisor function $\tau_k(n)$ can be represented in terms of the object distribution counting function N(n, m) by the following equation:*

$$\tau_k(n) = \sum_{r=0}^{k} \binom{k+1}{r} N(n, k-r+1). \tag{10.6}$$

Proof. Theorem 10.2 states that $N(n, m) = \sum_{k=0}^{m-1} (-1)^k \binom{m}{k} \tau_{m-k-1}(n)$. We can invert this using the Binomial inversion formula which states that if $b_n = \sum_{r=0}^{n} (-1)^r \binom{n}{r} a_{n-r}$, then $a_n = \sum_{r=0}^{n} \binom{n}{r} b_r$. After a change of variables and some algebra we obtain (10.6). □

Properties of the Generalized Divisor Function

The derivation and proof of Theorem 10.2 led naturally to a kind of generalized divisor function, $\tau_k(n)$. This function, $\tau_k(n)$, is called the *generalized divisor function* because $\tau_1(n) = \tau(n)$ is the well known *divisor function* from number theory. Basically, $\tau(n)$ counts the number of positive integer divisors of a positive integer n. The first few values of $\tau(n)$ are $\tau(1) = 1$, $\tau(2) = 2$, $\tau(3) = 2$, $\tau(4) = 3$, and so on. In Theorem 10.2, we defined the generalized divisor function recursively as $\tau_0(n) = 1$ and $\tau_{k+1}(n) = \sum_{d|n} \tau_k(d)$, where we sum over the divisors d of n. The recursive definition is sufficient to systematically calculate a table of values for $\tau_k(n)$ as provided in Table 10.1.

The generalized divisor function $\tau_k(n)$ has some remarkable mathematical properties that we will now state and prove as theorems.

Theorem 10.6. *The generalized divisor function, $\tau_k(n)$, is a multiplicative number-theoretic function. That is, if m and n are two coprime positive integers, then $\tau_k(mn) = \tau_k(m)\tau_k(n)$.*

Proof. We will prove this by mathematical induction. For the induction basis, note that if m and n are coprime, then it is trivially true that $\tau_0(mn) = \tau_0(m)\tau_0(n)$ since $\tau_0(n) = 1$ for all positive integers n, by definition. Also, $\tau_1(n)$ is multiplicative, because $\tau_1(n) = \tau(n)$ and $\tau(n)$ is known to be multiplicative. We must now show that if $\tau_k(n)$ is multiplicative, then $\tau_{k+1}(n)$ must be multiplicative. Assume that $\tau_k(n)$ is multiplicative for all positive integers n. Let $(m, n) = 1$. That is, m and n are coprime positive integers. Let d_j be a divisor of m and let δ_j be a divisor of n. Then we have

$$\begin{aligned}
\tau_{k+1}(m)\tau_{k+1}(n) &= \left(\sum_{d|m}\tau_k(d)\right)\left(\sum_{\delta|n}\tau_k(\delta)\right) \\
&= (\tau_k(1) + \tau_k(d_1) + \tau_k(d_2) + \ldots + \tau_k(m))(\tau_k(1) + \tau_k(\delta_1) + \tau_k(\delta_2) + \ldots + \tau_k(n)) \\
&= (1 + \tau_k(d_1) + \tau_k(d_2) + \ldots + \tau_k(m))(1 + \tau_k(\delta_1) + \tau_k(\delta_2) + \ldots + \tau_k(n)) \\
&= 1 + \tau_k(d_1) + \tau_k(d_2) + \ldots + \tau_k(\delta_1) + \tau_k(\delta_2) + \ldots + \tau_k(d_1)\tau_k(\delta_1) + \tau_k(d_1)\tau_k(\delta_2) + \ldots + \tau_k(m)\tau_k(n) \\
&= 1 + \tau_k(d_1) + \tau_k(d_2) + \ldots + \tau_k(\delta_1) + \tau_k(\delta_2) + \ldots + \tau_k(d_1\delta_1) + \tau_k(d_1\delta_2) + \ldots + \tau_k(mn) \\
&= \sum_{d|mn}\tau_k(mn) \\
&= \tau_{k+1}(mn) \quad \square
\end{aligned}$$

Example 10.7. Using the values provided in Table 10.1, we note, for example, that $\tau_2(6) = 9$. Also, $\tau_2(6) = \tau_2(2 \cdot 3) = \tau_2(2)\tau_2(3) = 3 \cdot 3 = 9$.

Theorem 10.6 is very useful for calculating values of $\tau_k(n)$ because any positive integer n, greater than 1, can be uniquely prime factorized. And if we know the values of $\tau_k(p^r)$ for all primes p, then we can calculate any value of $\tau_k(n)$.

Theorem 10.8. *For prime p,* $\tau_k(p^m) = \binom{m+k}{k}$.

Proof. By Lemma 10.5, we have $\tau_k(n) = \sum_{r=0}^{k}\binom{k+1}{r}N(n, k-r+1)$. For $n = p^m$, prime p, this becomes

$$\tau_k(p^m) = \sum_{r=0}^{k} \binom{k+1}{r} N(p^m, k-r+1). \tag{10.7}$$

What is $N(p^m, k-r+1)$? The quantity $N(p^m, k-r+1)$ is the number of ways to distribute objects of specification $n = p^m$, or m *identical objects* into precisely $k-r+1$ *distinct boxes* with no box empty. The number of ways to distribute a identical objects into precisely b labeled boxes is the same as the number of solutions to the equation $x_1 + x_2 + ... + x_b = a$ in positive integers $x_j \geq 1$. And that is well-known to be $\binom{a-1}{b-1}$. Therefore, $N(p^m, k-r+1) = \binom{m-1}{k-r}$. Substituting this into equation (10.7), we obtain

$$\tau_k(p^m) = \sum_{r=0}^{k} \binom{k+1}{r} \binom{m-1}{k-r} \tag{10.8}$$

We must now evaluate the combinatorial sum in (10.8). Vandermonde's identity states that

$$\sum_{r=0}^{k} \binom{a}{r} \binom{b}{k-r} = \binom{a+b}{k} \tag{10.9}$$

Setting $a = k+1$ and $b = m-1$, equation (10.8) becomes

$$\tau_k(p^m) = \sum_{r=0}^{k} \binom{k+1}{r} \binom{m-1}{k-r}$$
$$= \binom{m+k}{k} \quad \square$$

Theorems 10.6 and 10.8, taken together, allow us to easily calculate any value of $\tau_k(n)$.

We are now ready to state and prove a beautiful and remarkable theorem:

Theorem 10.9 (Partition-Convolution Identity). *Let $\tau_k(n)$ be the generalized divisor function defined by $\tau_{k+1}(n) = \sum_{d|n} \tau_k(n)$ with $\tau_0(n) = 1$. For any two integer partitions of k into precisely m parts, say $r_1 + r_2 + ... + r_m = k$ and $s_1 + s_2 + ... + s_m = k$, the following identity holds true:*

$$\tau_{r_1}(n) * \tau_{r_2}(n) * ... * \tau_{r_m}(n) = \tau_{s_1}(n) * \tau_{s_2}(n) * ... * \tau_{s_m}(n) \tag{10.10}$$

where $*$ is the standard number-theoretic convolution product.

Proof. Let $T(x) = \sum_{n=1}^{\infty} \tau(n) x^{\log n}$ be a logarithmic generating function for $\tau(n)$. We would like to find a logarithmic generating function for $\tau_k(n)$. To do this, let $U(x) = \sum_{n=1}^{\infty} x^{\log n}$ be a logarithmic generating function for the sequence $\{1, 1, ... \}$. Since multiplication of two logarithmic generating functions produces a convolution product, we have

$$T(x)U(x) = \left(\sum_{n=1}^{\infty} \tau(n) x^{\log n} \right) \left(\sum_{n=1}^{\infty} 1 x^{\log n} \right)$$

$$= \sum_{n=1}^{\infty} \left(\sum_{d|n} \tau(d) \right) x^{\log n}$$

$$= \sum_{n=1}^{\infty} \tau_2(n) x^{\log n}$$

Thus, $T(x)U(x)$ generates the sequence $\{\tau_2(n)\}$. By repeating this method, we can easily show that, in general, $T(x)U^k(x)$ generates the sequence $\{\tau_{k+1}(n)\}$. Hence, if $T_k(x)$ generates $\tau_k(n)$, then $T_1(x)U^k(x) = T_{k+1}(x)$ where $T_1(x) = T(x)$. Solving for $U^k(x)$, we get

$$U^k(x) = \frac{T_{k+1}(x)}{T_1(x)} \tag{10.11}$$

Now comes the insightful observation. On the left-hand side of (10.11), we can partition $U^k(x)$ as a product of powers of $U(x)$ in any way we want. For example, corresponding to the integer partition $7 = 3 + 2 + 2$, we have $U^7(x) = U^3(x)U^2(x)U^2(x)$. If we partition $U^k(x)$, that is, partition k, into, say, m parts in two different ways, then, from (10.11), we get

$$U^k(x) = U^k(x)$$

$$\frac{T_{r_1+1}(x)}{T_1(x)} \cdot \frac{T_{r_2+1}(x)}{T_1(x)} \cdot \ldots \cdot \frac{T_{r_m+1}(x)}{T_1(x)} = \frac{T_{s_1+1}(x)}{T_1(x)} \cdot \frac{T_{s_2+1}(x)}{T_1(x)} \cdot \ldots \cdot \frac{T_{s_m+1}(x)}{T_1(x)} \qquad (10.12)$$

$$T_{r_1+1}(x) \cdot T_{r_2+1}(x) \cdot \ldots \cdot T_{r_m+1}(x) = T_{s_1+1}(x) \cdot T_{s_2+1}(x) \cdot \ldots \cdot T_{s_m+1}(x)$$

The reason that we can cancel out all of the $T_1(x)$ functions is that we have the same number of them, m, on each side of the equation in (10.12). Since each subscript in (10.12) is incremented by 1, we can replace $r_j + 1$ by r_j and replace $s_j + 1$ by s_j (just a change of variables to simplify the notation) to get

$$T_{r_1}(x) \cdot T_{r_2}(x) \cdot \ldots \cdot T_{r_m}(x) = T_{s_1}(x) \cdot T_{s_2}(x) \cdot \ldots \cdot T_{s_m}(x) \qquad (10.13)$$

Finally, since multiplication of logarithmic generating functions corresponds to convolution products of their number sequences, we obtain

$$\tau_{r_1}(n) * \tau_{r_2}(n) * \ldots * \tau_{r_m}(n) = \tau_{s_1}(n) * \tau_{s_2}(n) * \ldots * \tau_{s_m}(n) \quad \square \qquad (10.14)$$

Corollary 10.10. *Let* $k_1 + k_2 = r_1 + r_2$ *for positive integers* k_j *and* r_j. *Then*

$$\sum_{d|n} \tau_{k_1}(d) \tau_{k_2}(n/d) = \sum_{d|n} \tau_{r_1}(d) \tau_{r_2}(n/d).$$

Proof. Let $m = 2$ in Theorem 10.9.

Example 10.11. Let $n = 14$. Consider the partitions of 5 into two parts: $3 + 2$ and $4 + 1$. By Corollary 10.10, we have $\sum_{d|14} \tau_3(d) \tau_2(14/d) = \sum_{d|14} \tau_4(d) \tau_1(14/d)$. These two quantities are equal as we observe:

$$\sum_{d|14} \tau_3(d) \tau_2(14/d) = \tau_3(1)\tau_2(14) + \tau_3(2)\tau_2(7) + \tau_3(7)\tau_2(2) + \tau_3(14)\tau_2(1)$$

$$= 1 \cdot 9 + 4 \cdot 3 + 4 \cdot 3 + 16 \cdot 1$$

$$= 49$$

Also, we have

$$\sum_{d|14}\tau_4(d)\tau_1(14/d) = \tau_4(1)\tau_1(14)+\tau_4(2)\tau_1(7)+\tau_4(7)\tau_1(2)+\tau_4(14)\tau_1(1)$$
$$= 1\cdot 4+5\cdot 2+5\cdot 2+25\cdot 1$$
$$= 49$$

As a special case, Theorem 10.2 gives us a well-known identity with little additional effort:

Theorem 10.12. $\sum_{k=0}^{m-1}(-1)^k\binom{m}{k}(m-k)^m = m!$

Proof. Let the specification for objects, n, be a product of m distinct primes. Using Theorems 10.2, 10.6, and 10.8, we have

$$N(p_1, p_2, \ldots, p_m, m) = \sum_{k=0}^{m-1}(-1)^k\binom{m}{k}\tau_{m-k-1}(p_1 p_2 \cdots p_m)$$
$$= \sum_{k=0}^{m-1}(-1)^k\binom{m}{k}\tau_{m-k-1}(p_1)\tau_{m-k-1}(p_2)\cdots\tau_{m-k-1}(p_m)$$
$$= \sum_{k=0}^{m-1}(-1)^k\binom{m}{k}\binom{m-k}{m-k-1}\binom{m-k}{m-k-1}\cdots\binom{m-k}{m-k-1}$$
$$= \sum_{k=0}^{m-1}(-1)^k\binom{m}{k}(m-k)^m$$

On the other hand, what is $N(p_1 p_2 \cdots p_m, m)$? In fact, $N(p_1 p_2 \cdots p_m, m) = m!$ since m distinct objects can be placed into m distinct boxes, with no box empty, in $m!$ ways. □

Using Theorems 10.2, 10.6, and 10.8, we can, if desired, put everything together in a nice package. Suppose, for example, that we have a red balls, b white balls, and c blue balls. Then the number of ways to distribute these colored balls into precisely m distinct boxes, with no box empty, is

$$N((a, b, c), m) = \sum_{k=0}^{m-1}(-1)^k\binom{m}{k}\binom{a+m-k-1}{a}\binom{b+m-k-1}{b}\binom{c+m-k-1}{c}$$

The generalization is obvious. It is amazing how much we can squeeze out of a single generating function like $\sum_{n=1}^{\infty}g_n(t, \alpha_1, \alpha_2, \ldots, \alpha_m)x^{\log n} = \prod_{d=2}^{\infty}\dfrac{1}{(1-t\alpha_1 x^{\log d})(1-t\alpha_2 x^{\log d})\cdots(1-t\alpha_m x^{\log d})}$.

Table 10.1. Table of Generalized Divisor Function Values

n	$\tau_0(n)$	$\tau_1(n)$	$\tau_2(n)$	$\tau_3(n)$	$\tau_4(n)$	$\tau_5(n)$
1	1	1	1	1	1	1
2	1	2	3	4	5	6
3	1	2	3	4	5	6
4	1	3	6	10	15	21
5	1	2	3	4	5	6
6	1	4	9	16	25	36
7	1	2	3	4	5	6
8	1	4	10	20	35	56
9	1	3	6	10	15	21
10	1	4	9	16	25	36
11	1	2	3	4	5	6
12	1	6	18	40	75	126
13	1	2	3	4	5	6
14	1	4	9	16	25	36
15	1	4	9	16	25	36
16	1	5	15	35	70	126
17	1	2	3	4	5	6
18	1	6	18	40	75	126
19	1	2	3	4	5	6
20	1	6	18	40	75	126
21	1	4	9	16	25	36
22	1	4	9	16	25	36
23	1	2	3	4	5	6
24	1	8	30	80	175	336
25	1	3	6	10	15	21
26	1	4	9	16	25	36
27	1	4	10	20	35	56
28	1	6	18	40	75	126
29	1	2	3	4	5	6
30	1	8	27	64	125	216
31	1	2	3	4	5	6
32	1	6	21	56	126	252
33	1	4	9	16	25	36
34	1	4	9	16	25	36
35	1	4	9	16	25	36
36	1	9	36	100	225	441

Given a set of *arbitrary* objects, such as colored balls, in how many ways can we distribute the objects into *distinct* boxes with no box empty? In the previous sections, we developed one approach to this problem. Another way to solve this problem is to build a generating function. The generating function will allow us to calculate object distribution polynomials, $h_n(t)$, that solve the enumeration problem. The coefficient of $t^k/k!$ in $h_n(t)$ is the number of ways to distribute objects, of specification n, into precisely k distinct, or labeled, boxes (cells). Using the number-theoretic paradigm developed in this book, we can build a logarithmic generating function that allows us to calculate the object distribution polynomials $h_n(t)$.

Generating Function for Distributing Objects into Distinct Boxes

How can we build a generating function to calculate the object distribution polynomial $h_n(t)$? We are distributing objects of arbitrary specification number n into distinct, or labeled, boxes. Since the boxes are labeled, we must take permutations into account. The number of permutations of k things, a of which are of one kind, b of another kind, and so on, is given by $\dfrac{k!}{a!b!...}$, which is the coefficient of $\dfrac{t^k}{k!}$ in the product $\dfrac{t^a}{a!}\dfrac{t^b}{b!}...$, where $a+b+...=k$. So, the polynomial $h_n(t)$ is of *exponential type*

$$h_n(t) = \sum_{k \geq 1} h(n,k) \frac{t^k}{k!}. \qquad (10.15)$$

We want to build a generating function $H(x,t)$ such that

$$H(x,t) = \sum_{n=1}^{\infty} h_n(t) x^{\log n}. \qquad (10.16)$$

Let's consider a simple example. Suppose that we have one red ball and one blue ball. We can represent this set of objects (colored balls) by the set $S = \{r, b\} = \{2, 3\}$, where a red ball is represented by the prime number 2 and a blue ball is represented by the prime 3. To find the specification number of the set of objects, we let $n = 2 \cdot 3 = 6$. Now consider the generating function $\left(1 + \dfrac{t}{1!} x^{\log 2}\right)\left(1 + \dfrac{t}{1!} x^{\log 3}\right)\left(1 + \dfrac{t}{1!} x^{\log 6}\right)$. The generating function includes a term for each non-unit divisor of 6, namely, 2, 3, and 6. If we expand this generating function, by multiplying out the product, and look only at the coefficient of $x^{\log 6}$, we get the term $\left(\dfrac{t}{1!} + 2\dfrac{t^2}{2!}\right) x^{\log 6}$. This tells us that

$h_6(t) = \dfrac{t}{1!} + 2\dfrac{t^2}{2!}$. For example, since the coefficient of $t^2/2!$ is 2, there are two ways to distribute one red ball and one blue ball into precisely two distinct boxes with no box empty:

This method generalizes for all specification numbers n:

$$\sum_{n=1}^{\infty} h_n(t) x^{\log n} = \left(1 + \dfrac{t}{1!}x^{\log 2} + \dfrac{t^2}{2!}x^{\log 2^2} + \ldots\right)\left(1 + \dfrac{t}{1!}x^{\log 3} + \dfrac{t^2}{2!}x^{\log 3^2} + \ldots\right)\left(1 + \dfrac{t}{1!}x^{\log 4} + \dfrac{t^2}{2!}x^{\log 4^2} + \ldots\right)\ldots$$

This gives us the following theorem which simply formalizes the preceding construction in the language of mathematics:

Theorem 10.13. *Let $h_n(t)$ be the exponential type object distribution polynomial that counts the number of ways to distribute objects of specification n into distinct (labeled) boxes with no box empty. A generating function for $h_n(t)$ is given by*

$$H(x,t) = \sum_{n=1}^{\infty} h_n(t) x^{\log n} = \prod_{d=2}^{\infty}\left(\sum_{k=0}^{\infty} \dfrac{t^k}{k!} x^{\log d^k}\right).$$

Theorem 10.14. *Let $h_n(t)$ be the exponential-type object distribution polynomial that counts the number of ways to distribute objects of specification n into distinct boxes with no box empty. Then*

$$H(x,t) = \sum_{n=1}^{\infty} h_n(t) x^{\log n} = \prod_{d=2}^{\infty} e^{t \, x^{\log d}}.$$

Proof. By Theorem 10.13, we have

$$\sum_{n=1}^{\infty} h_n(t) x^{\log n} = \prod_{d=2}^{\infty} \left(\sum_{k=0}^{\infty} \frac{t^k}{k!} x^{\log d^k} \right)$$

$$= \prod_{d=2}^{\infty} \left(\sum_{k=0}^{\infty} \frac{t^k}{k!} x^{k \log d} \right)$$

$$= \prod_{d=2}^{\infty} \left(\sum_{k=0}^{\infty} \frac{(t x^{\log d})^k}{k!} \right)$$

$$= \prod_{d=2}^{\infty} e^{t x^{\log d}}. \quad \square$$

Theorem 10.14 can also be represented as a Dirichlet series generating function as follows.

Theorem 10.15. *Let* $h_n(t)$ *be the exponential-type object distribution polynomial that counts the number of ways to distribute objects of specification n into distinct boxes with no box empty. Then a Dirichlet series generating function for* $h_n(t)$ *is given by*

$$H(s, t) = \sum_{n=1}^{\infty} \frac{h_n(t)}{n^s} = \prod_{d=2}^{\infty} e^{t/d^s}.$$

Proof. Let $x = e^{-s}$ in Theorem 10.14. Then

$$\sum_{n=1}^{\infty} h_n(t)(e^{-s})^{\log n} = \prod_{d=2}^{\infty} e^{t(e^{-s})^{\log d}}$$

$$\sum_{n=1}^{\infty} h_n(t) e^{-s \log n} = \prod_{d=2}^{\infty} e^{t e^{-s \log d}}$$

$$\sum_{n=1}^{\infty} h_n(t) e^{\log n^{-s}} = \prod_{d=2}^{\infty} e^{t e^{\log d^{-s}}}$$

$$\sum_{n=1}^{\infty} h_n(t) n^{-s} = \prod_{d=2}^{\infty} e^{t d^{-s}}. \quad \square$$

Example 10.16. Show that $\sum_{n=1}^{\infty} \frac{h_n(1)}{n^2} = e^{\frac{\pi^2}{6} - 1}$.

Solution. By Theorem 10.14, $\sum_{n=1}^{\infty} h_n(t) x^{\log n} = \prod_{d=2}^{\infty} e^{tx^{\log d}}$. If we let $t=1$, then we get

$\sum_{n=1}^{\infty} h_n(1) x^{\log n} = \prod_{d=2}^{\infty} e^{x^{\log d}}$. Now let $x = e^{-2}$. Then $\sum_{n=1}^{\infty} h_n(1) e^{-2\log n} = \prod_{d=2}^{\infty} e^{e^{-2\log d}}$, or $\sum_{n=1}^{\infty} \frac{h_n(1)}{n^2} = \prod_{d=2}^{\infty} e^{1/d^2} = k$.

Then $\log k = \log\left(\prod_{d=2}^{\infty} e^{1/d^2}\right) = \sum_{d=2}^{\infty} \log e^{1/d^2} = \sum_{d=2}^{\infty} \frac{1}{d^2} = \frac{\pi^2}{6} - 1$. Therefore, $k = e^{\frac{\pi^2}{6}-1}$. So,

$\sum_{n=1}^{\infty} \frac{h_n(1)}{n^2} = e^{\frac{\pi^2}{6}-1} \approx 1.90586.$

If we let $h(n, k)$ denote the number of ways to distribute objects of specification n into precisely k distinct boxes with no box empty, then the object distribution polynomial $h_n(t)$ can be written as

$$h_n(t) = \sum_{k=1}^{\infty} h(n, k) \frac{t^k}{k!}. \tag{10.17}$$

The quantity $h(n, k)$ is also the number of ordered factorizations of the positive integer n into precisely k non-unit factors. By definition, we take $h_1(t) = 1$. Using (10.17), we can represent Theorem 10.14 in the following form:

Theorem 10.17. $\sum_{n=1}^{\infty} \sum_{k=1}^{\infty} h(n, k) \frac{t^k}{k!} x^{\log n} = \prod_{d=2}^{\infty} e^{tx^{\log d}}.$

Proof. Let $h_n(t) = \sum_{k=1}^{\infty} h(n, k) \frac{t^k}{k!}$ in Theorem 10.14. □

Lemma 10.18. $\frac{d}{dt} h_n(t) = \sum_{d|n} h_d(t) - h_n(t) = \sum_{\substack{d|n \\ d \neq n}} h_d(t).$

Proof. By Theorem 10.14, we have $\sum_{n=1}^{\infty} h_n(t) x^{\log n} = \prod_{d=2}^{\infty} e^{tx^{\log d}}$. Then

$$\log\left(\sum_{n=1}^{\infty} h_n(t) x^{\log n}\right) = \log\left(\prod_{d=2}^{\infty} e^{t x^{\log d}}\right)$$

$$= \sum_{d=2}^{\infty} \log e^{t x^{\log d}}$$

$$= \sum_{d=2}^{\infty} t x^{\log d}.$$

Taking derivatives with respect to t, we get

$$\frac{d}{dt} \log\left(\sum_{n=1}^{\infty} h_n(t) x^{\log n}\right) = \frac{d}{dt} \sum_{d=2}^{\infty} t x^{\log d}$$

$$\frac{\frac{d}{dt}\sum_{n=1}^{\infty} h_n(t) x^{\log n}}{\sum_{n=1}^{\infty} h_n(t) x^{\log n}} = \sum_{d=2}^{\infty} x^{\log d}$$

$$\sum_{n=1}^{\infty} \frac{d}{dt} h_n(t) x^{\log n} = \sum_{n=1}^{\infty} h_n(t) x^{\log n} \cdot \sum_{d=2}^{\infty} x^{\log d}$$

$$= \sum_{n=1}^{\infty} h_n(t) x^{\log n} \left(\sum_{d=1}^{\infty} x^{\log d} - 1\right)$$

$$= \sum_{n=1}^{\infty} h_n(t) x^{\log n} \cdot \sum_{n=1}^{\infty} x^{\log n} - \sum_{n=1}^{\infty} h_n(t) x^{\log n}$$

$$= \sum_{n=1}^{\infty} (h_n(t) * 1) x^{\log n} - \sum_{n=1}^{\infty} h_n(t) x^{\log n}$$

$$= \sum_{n=1}^{\infty} \left(\sum_{d|n} h_d(t) \cdot 1\right) x^{\log n} - \sum_{n=1}^{\infty} h_n(t) x^{\log n}.$$

Equating coefficients of $x^{\log n}$ gives $\dfrac{d}{dt} h_n(t) = \sum_{d|n} h_d(t) - h_n(t)$. $\square$

We are now prepared to derive an integral recurrence formula for calculating the object distribution polynomials $h_n(t)$. Explicit formulas for $h_n(t)$ are known only for special cases of the object sets. Nevertheless, we can use the methods developed in this section to derive a recurrence relation for $h_n(t)$. This will allow us to actually calculate the polynomials $h_n(t)$.

Theorem 10.19. Let $h_n(t)$ be the exponential-type object distribution polynomial that counts the number of ways to distribute objects of specification n into distinct boxes with no box empty. A recurrence relation for $h_n(t)$ is given by

$$h_n(t) = \sum_{\substack{d|n \\ d \neq n}} \int_0^t h_d(u)\,du,$$

where $h_1(t) = 1$.

Proof. By Lemma 10.18, we have $\dfrac{d}{dt} h_n(t) = \sum_{\substack{d|n \\ d \neq n}} h_d(t)$. Integrating this equation gives

$$h_n(t) = \int_0^t \sum_{\substack{d|n \\ d \neq n}} h_d(u)\,du$$

$$= \sum_{\substack{d|n \\ d \neq n}} \int_0^t h_d(u)\,du. \quad \square$$

The recurrence relation in Theorem 10.19 can be used to calculate a reference table of polynomials $h_n(t)$. Table 10.2 gives several of these polynomials.

Table 10.2. Object Distribution polynomials $h_n(t)$

Specification Number n	$h_n(t)$
1	$h_1(t) = 1$
2	$h_2(t) = \dfrac{t}{1!}$
3	$h_3(t) = \dfrac{t}{1!}$
4	$h_4(t) = \dfrac{t}{1!} + \dfrac{t^2}{2!}$
5	$h_5(t) = \dfrac{t}{1!}$
6	$h_6(t) = \dfrac{t}{1!} + 2\dfrac{t^2}{2!}$
7	$h_7(t) = \dfrac{t}{1!}$
8	$h_8(t) = \dfrac{t}{1!} + 2\dfrac{t^2}{2!} + \dfrac{t^3}{3!}$
9	$h_9(t) = \dfrac{t}{1!} + \dfrac{t^2}{2!}$
10	$h_{10}(t) = \dfrac{t}{1!} + 2\dfrac{t^2}{2!}$
11	$h_{11}(t) = \dfrac{t}{1!}$
12	$h_{12}(t) = \dfrac{t}{1!} + 4\dfrac{t^2}{2!} + 3\dfrac{t^3}{3!}$

Theorem 10.19 can be reinterpreted as a statement about the solutions to a *functional equation*.

Theorem 10.20. *The solutions $h_n(t)$ to the functional integral equation*

$$h_n(t) = \sum_{\substack{d \mid n \\ d \neq n}} \int_0^t h_d(u)\, du$$

with $h_1(t) = 1$, are the exponential type object distribution polynomials that count the number of ways to distribute objects of specification n into distinct boxes with no box empty.

Example 10.21. Suppose we have two red balls and one blue ball. The multiset of objects is given by $S = \{r, r, b\}$. If we represent a red ball by the prime number 2 and a blue ball by the prime number 3, then $S = \{2, 2, 3\}$. A specification number for this set of objects is $n = 2 \cdot 2 \cdot 3 = 12$. We can use the integral recurrence formula in Theorem 10.19 to calculate the object distribution polynomial $h_{12}(t)$ which counts the number of ways to distribute two red balls and one blue ball into distinct, or labeled, boxes with no box empty. Once we calculate $h_{12}(t)$, the coefficient of $t^k/k!$ gives the number of ways to distribute two red balls and one blue ball into precisely k distinct boxes with no box empty. The divisors of 12 are $\{1, 2, 3, 4, 6, 12\}$. So, by Theorem 10.19 and Table 10.2, we have

$$h_{12}(t) = \sum_{\substack{d|12 \\ d \neq 12}} \int_0^t h_d(u)\,du$$

$$= \int_0^t h_1(u)\,du + \int_0^t h_2(u)\,du + \int_0^t h_3(u)\,du + \int_0^t h_4(u)\,du + \int_0^t h_6(u)\,du$$

$$= \int_0^t 1\,du + \int_0^t u\,du + \int_0^t u\,du + \int_0^t \left(u + \frac{u^2}{2!}\right)du + \int_0^t \left(u + 2\frac{u^2}{2!}\right)du$$

$$= \frac{t}{1!} + 4\frac{t^2}{2!} + 3\frac{t^3}{3!}.$$

Theorem 10.22. *Let $h(n, k)$ denote the number of ways to distribute objects of specification n into precisely k distinct boxes with no box empty. (We defined $h(n, k)$ in equation (10.15).) A recurrence relation for $h(n, k)$ is given by*

$$h(n, k) = \sum_{\substack{d|n \\ d \neq n}} h(d, k-1),$$

where $h(1, k) = 0$ for $k > 1$ and $h(1, 1) = 0$.

Proof. By Theorem 10.19, we have

$$h_n(t) = \sum_{\substack{d|n \\ d \neq n}} \int_0^t h_d(u)\, du$$

$$= \sum_{\substack{d|n \\ d \neq n}} \int_0^t \sum_{k \geq 1} h(d, k) \frac{u^k}{k!}\, du$$

$$= \sum_{\substack{d|n \\ d \neq n}} \sum_{k \geq 1} \int_0^t h(d, k) \frac{u^k}{k!}\, du$$

$$= \sum_{\substack{d|n \\ d \neq n}} \sum_{k \geq 1} h(d, k) \frac{t^{k+1}}{(k+1)!}$$

$$= \sum_{k \geq 1} \left(\sum_{\substack{d|n \\ d \neq n}} h(d, k) \right) \frac{t^{k+1}}{(k+1)!}.$$

Also, $h_n(t) = \sum_{k \geq 1} h(n, k) \frac{t^k}{k!}$, so we have $\sum_{k \geq 1} h(n, k) \frac{t^k}{k!} = \sum_{k \geq 1} \left(\sum_{\substack{d|n \\ d \neq n}} h(d, k) \right) \frac{t^{k+1}}{(k+1)!}$. Equating coefficients of $\frac{t^k}{k!}$ gives $h(n, k) = \sum_{\substack{d|n \\ d \neq n}} h(d, k-1)$. □

Caution: The number n is *not* the number of objects. Rather, n is the *specification number* that encodes the object set according to Definition 2.1.

The polynomials $h_n(t)$ satisfy a simple convolution identity relationship:

Theorem 10.23. $h_n(t) * h_n(-t) = \begin{cases} 1, & n = 1; \\ 0, & n \neq 1. \end{cases}$

Proof. By Theorem 10.14, we have

$$H(x,t) \cdot H(x,-t) = \sum_{n=1}^{\infty} h_n(t) x^{\log n} \cdot \sum_{n=1}^{\infty} h_n(-t) x^{\log n}$$

$$= \prod_{d=2}^{\infty} e^{tx^{\log d}} \cdot \prod_{d=2}^{\infty} e^{-tx^{\log d}}$$

$$= \left(e^{tx^{\log 2}} e^{tx^{\log 3}} e^{tx^{\log 4}} \ldots \right)\left(e^{-tx^{\log 2}} e^{-tx^{\log 3}} e^{-tx^{\log 4}} \ldots \right)$$

$$= \left(e^{tx^{\log 2}} e^{-tx^{\log 2}} \right)\left(e^{tx^{\log 3}} e^{-tx^{\log 3}} \right)\left(e^{tx^{\log 4}} e^{-tx^{\log 4}} \right)\ldots$$

$$= e^0 e^0 e^0 \ldots$$

$$= 1.$$

Also, $\sum_{n=1}^{\infty} h_n(t) x^{\log n} \cdot \sum_{n=1}^{\infty} h_n(-t) x^{\log n} = \sum_{n=1}^{\infty} h_n(t) * h_n(-t) x^{\log n}$. So we have $\sum_{n=1}^{\infty} h_n(t) * h_n(-t) x^{\log n} = 1 = 1 x^{\log 1}$.

Comparing coefficients of $x^{\log n}$ gives

$$h_n(t) * h_n(-t) = \begin{cases} 1, & n=1; \\ 0, & n \neq 1. \end{cases} \quad \square$$

Example 10.24. Let $n = 4$. We have

$$h_4(t) * h_4(-t) = \sum_{d|4} h_d(t) h_{\frac{4}{d}}(-t)$$

$$= h_1(t) h_4(-t) + h_2(t) h_2(-t) + h_4(t) h_1(-t)$$

$$= 1 \cdot \left(-t + \frac{t^2}{2!} \right) + (t)(-t) + \left(t + \frac{t^2}{2!} \right) \cdot 1$$

$$= 0.$$

Generating Function for Distributing Objects into Distinct Boxes

Special cases of Theorem 10.19 give us well known combinatorial results. Theorem 10.25 is one example. There are much easier ways to prove the formula given in Theorem 10.25, but we want to show that Theorem 10.19, which is a very general result, gives the correct answer for special cases.

Theorem 10.25. *For $r \geq k$, the number of ways to distribute r identical objects into k distinct boxes with no box empty is given by*

$$h(p^r, k) = \binom{r-1}{k-1}.$$

Note that r identical objects will have a specification number $n = p^r$ for some prime p.

Proof. Let $n = p^r$. By Theorem 10.19, we have $h_{p^r}(t) = \sum_{\substack{d \mid p^r \\ d \neq p^r}} \int_0^t h_d(u)\,du$, where $h_1(t) = 1$. Then, expanding this recurrence relation gives

$$h_{p^r}(t) = \sum_{\substack{d \mid p^r \\ d \neq p^r}} \int_0^t h_d(u)\,du$$

$$= \int_0^t h_1(u)\,du + \int_0^t h_p(u)\,du + \ldots + \int_0^t h_{p^{r-1}}(u)\,du$$

$$= \int_0^t h_1(u)\,du + \int_0^t \sum_{\substack{d \mid p \\ d \neq p}} \int_0^t h_d(u)\,du\,du + \ldots + \int_0^t \sum_{\substack{d \mid p^{r-1} \\ d \neq p^{r-1}}} \int_0^t h_d(u)\,du\,du$$

$$= \int_0^t 1\,du + \binom{r-1}{1}\int_0^t \int_0^t 1\,du\,du + \binom{r-1}{2}\int_0^t \int_0^t \int_0^t 1\,du\,du\,du + \ldots + \binom{r-1}{r-1}\int_0^t \ldots \int_0^t 1\,du\ldots du$$

$$= \sum_{j=0}^{r-1} \binom{r-1}{j} \int_0^t \ldots \int_0^t 1\,du\ldots du$$

$$= \sum_{j=0}^{r-1} \binom{r-1}{j} \frac{t^{j+1}}{(j+1)!}.$$

Now let $k = j+1$ and we get $h_{p^r}(t) = \sum_{k=1}^{r} \binom{r-1}{k-1} \frac{t^k}{k!} = \sum_{k=1}^{r} h(p^r, k) \frac{t^k}{k!}$. Equating coefficients of $t^k/k!$ gives the final result: $h(p^r, k) = \binom{r-1}{k-1}$.

For notational convenience, let $f_r(t)$ be the exponential-type object distribution polynomial that counts the number of ways to distribute r identical objects into *distinct boxes* with no box empty. Thus, $f_r(t) = h_{p^r}(t)$. The following two theorems give another approach to evaluating $f_r(t) = h_{p^r}(t)$.

Theorem 10.26. *If $e^{tu/(1-u)}$ is expanded as an ordinary power series in the variable u, then the coefficient of u^r is the exponential object distribution polynomial $f_r(t)$ for counting the number of distributions of r identical objects into distinct boxes with no box empty.*

Proof. For r identical objects, $n = p^r$, for a prime number p. A generating function for $h_n(t)$ is given by $e^{tx^{\log p}} \cdot e^{tx^{\log p^2}} \cdot e^{tx^{\log p^3}} \cdots = e^{t(x^{\log p} + x^{2\log p} + x^{3\log p} + \cdots)}$. If we let $u = x^{\log p}$, then

$$x^{\log p} + x^{2\log p} + x^{3\log p} + \cdots = u + u^2 + u^3 + \cdots$$
$$= \frac{u}{1-u}.$$

So, our generating function for $f_r(t) = h_{p^r}(t)$ becomes $e^{tu/(1-u)}$. Then we have

$$e^{tu/(1-u)} = 1 + tu + \left(t + \frac{t^2}{2!}\right)u^2 + \left(t + 2\frac{t^2}{2!} + \frac{t^3}{3!}\right)u^3 + \cdots = \sum_{r=0}^{\infty} f_r(t)u^r.$$

Thus, if we expand $e^{tu/(1-u)}$ as an ordinary power series in the variable u, the coefficient of u^r is the exponential-type object distribution polynomial $f_r(t) = h_{p^r}(t)$. If we let $f(r, k)$ be the number of ways to distribute r identical objects into *precisely* k distinct boxes with no box empty, then

$$\sum_{k=1}^{\infty} f(r, k) \frac{t^k}{k!} = [u^r] e^{tu/(1-u)}$$
$$= \text{the coefficient of } u^r \text{ in } e^{tu/(1-u)}. \quad \square$$

Theorem 10.27. *Let $f_r(t)$ be the exponential object distribution polynomial that counts the number of distributions of r identical objects into distinct boxes with no box empty. A recurrence relation for $f_r(t)$, with $f_0(t) = 1$ and $f_1(t) = t$, is given by*

$$f_r(t) = \frac{1}{r}\sum_{k=1}^{r} k t f_{r-k}(t).$$

Proof. By Theorem 10.26, we have $\sum_{r=0}^{\infty} f_r(t)u^r = e^{tu/(1-u)}$. Taking logarithms and then differentiating with respect to the variable u, we get

$$\log\left(\sum_{r=0}^{\infty} f_r(t) u^r\right) = \frac{tu}{1-u}$$

$$\frac{d}{du}\log\left(\sum_{r=0}^{\infty} f_r(t) u^r\right) = \frac{d}{du}\left(\frac{tu}{1-u}\right)$$

$$\frac{\sum_{r=0}^{\infty} r f_r(t) u^{r-1}}{\sum_{r=0}^{\infty} f_r(t) u^r} = \frac{t}{(1-u)^2}$$

$$\sum_{r=0}^{\infty} r f_r(t) u^{r-1} = \frac{t}{(1-u)^2} \sum_{r=0}^{\infty} f_r(t) u^r$$

$$\sum_{r=0}^{\infty} r f_r(t) u^r = \frac{tu}{(1-u)^2} \sum_{r=0}^{\infty} f_r(t) u^r$$

$$\sum_{r=0}^{\infty} r f_r(t) u^r = \sum_{r=0}^{\infty} r t u^r \sum_{r=0}^{\infty} f_r(t) u^r$$

$$\sum_{r=0}^{\infty} r f_r(t) u^r = \sum_{r=0}^{\infty} \left(\sum_{k=0}^{r} k t f_{r-k}(t)\right) u^r.$$

Comparing coefficients of u^r on both sides of the last equation gives

$$r f_r(t) = \sum_{k=0}^{r} k t f_{r-k}(t)$$

$$= \sum_{k=1}^{r} k t f_{r-k}(t). \quad \square$$

An Upper Summation Identity

The real value of the number-theoretic method is using it to solve difficult combinatorial distribution and occupancy problems. However, if the theory is really true, then special cases of the theorems should reduce to well-known combinatorial theorems. This is, indeed, what happens, and I now present one such example:

Theorem 10.28 (Upper Summation Identity). *For positive integers n and m with $n \geq m$, we have*

$$\sum_{j=m}^{n} \binom{j}{m} = \binom{n+1}{m+1}.$$

Proof. By Theorem 10.25, the number of ways to distribute r identical objects into k distinct boxes with no box empty is $h(p^r, k) = \binom{r-1}{k-1}$. So, we have, for $n = p^r$, p prime,

$$h_n(t) = \sum_{k=1}^{r} h(p^r, k) \frac{t^k}{k!} = \sum_{k=1}^{r} \binom{r-1}{k-1} \frac{t^k}{k!}.$$

Then, using Theorem 10.19, we have

$$\sum_{k=1}^{r} \binom{r-1}{k-1} \frac{t^k}{k!} = \sum_{\substack{d \mid p^r \\ d \neq p^r}} \int_0^t h_d(u)\, du$$

$$= \int_0^t h_{p^0}(u)\, du + \int_0^t h_{p^1}(u)\, du + \ldots + \int_0^t h_{p^{r-1}}(u)\, du$$

$$= \int_0^t 1\, du + \int_0^t \sum_{k=1}^{1} h(p^1, k) \frac{u^k}{k!}\, du + \ldots + \int_0^t \sum_{k=1}^{r-1} h(p^{r-1}, k) \frac{u^k}{k!}\, du$$

$$= t + \int_0^t \sum_{k=1}^{1} \binom{1-1}{k-1} \frac{u^k}{k!}\, du + \ldots + \int_0^t \sum_{k=1}^{r-1} \binom{(r-1)-1}{k-1} \frac{u^k}{k!}\, du$$

$$= t + \sum_{k=1}^{1} \binom{1-1}{k-1} \frac{t^{k+1}}{(k+1)!} + \ldots + \sum_{k=1}^{r-1} \binom{(r-1)-1}{k-1} \frac{t^{k+1}}{(k+1)!}$$

$$= t + \left(\binom{0}{0} + \binom{1}{0} + \ldots + \binom{r-2}{0}\right) \frac{t^2}{2!} + \left(\binom{1}{1} + \binom{2}{1} + \ldots + \binom{r-2}{1}\right) \frac{t^3}{3!} + \ldots$$

$$+ \binom{r-2}{r-2} \frac{t^r}{r!}.$$

Equating coefficients of $t^k/k!$ gives $\sum_{j=m}^{n} \binom{j}{m} = \binom{n+1}{m+1}$. $\square$

Distinct Objects in Distinct Boxes

For r distinct objects, the specification number is $n = p_1 p_2 \ldots p_r$, where the p_i are distinct primes. There is a simple formula for counting the number of ways to distribute r distinct objects into k distinct boxes, for $r \geq k$: $\left\{{r \atop k}\right\} k!$, where $\left\{{r \atop k}\right\}$ are Stirling numbers of the second kind, or the number of ways to partition an r-set in k non-empty disjoint subsets. We can substitute this formula into equation (10.17) to obtain the following identity that relates binomial coefficients $\binom{n}{j}$ and Stirling numbers of the second kind $\left\{\begin{array}{c}n\\j\end{array}\right\}$.

Theorem 10.29. For positive integers n and m with $n \geq m$,

$$\frac{1}{m+1}\sum_{j=m}^{n}\binom{n+1}{j}\left\{\begin{array}{c}j\\m\end{array}\right\}=\left\{\begin{array}{c}n+1\\m+1\end{array}\right\}.$$

Proof. Let $n = p_1 p_2 \ldots p_r$. Then, from the discussion above, we have

$$h_{p_1 p_2 \ldots p_r}(t) = \sum_{k=1}^{r} h(p_1 p_2 \ldots p_r, k)\frac{t^k}{k!}$$

$$= \sum_{k=1}^{r}\left\{\begin{array}{c}r\\k\end{array}\right\}k!\frac{t^k}{k!}$$

$$= \sum_{k=1}^{r}\left\{\begin{array}{c}r\\k\end{array}\right\}t^k.$$

Also, by Theorem 10.19, we have

$$h_{p_1 p_2 \ldots p_r}(t) = \sum_{\substack{d \mid p_1 p_2 \ldots p_r \\ d \neq p_1 p_2 \ldots p_r}} \int_0^t h_d(u)\,du$$

$$= \binom{r}{0}\int_0^t 1\,du + \binom{r}{1}\int_0^t h_{p_1}(u)\,du + \ldots + \binom{r}{r-1}\int_0^t h_{p_1 p_2 \ldots p_{r-1}}(u)\,du$$

$$= \binom{r}{0}t + \binom{r}{1}\int_0^t \sum_{k=1}^{1}\left\{\begin{array}{c}1\\k\end{array}\right\}k!\frac{u^k}{k!}\,du + \ldots + \binom{r}{r-1}\int_0^t \sum_{k=1}^{r-1}\left\{\begin{array}{c}r-1\\k\end{array}\right\}k!\frac{u^k}{k!}\,du$$

$$= \binom{r}{0}t + \binom{r}{1}\left\{\begin{array}{c}1\\1\end{array}\right\}\frac{t^2}{2} + \binom{r}{2}\left(\left\{\begin{array}{c}2\\1\end{array}\right\}\frac{t^2}{2} + \left\{\begin{array}{c}2\\2\end{array}\right\}\frac{t^3}{3}\right) + \ldots + \binom{r}{r-1}\left(\left\{\begin{array}{c}r-1\\1\end{array}\right\}\frac{t^2}{2} + \left\{\begin{array}{c}r-1\\2\end{array}\right\}\frac{t^3}{3} + \ldots + \left\{\begin{array}{c}r-1\\r-1\end{array}\right\}\frac{t^r}{r}\right)$$

$$= \binom{r}{0}t + \left(\binom{r}{1}\left\{\begin{array}{c}1\\1\end{array}\right\} + \binom{r}{2}\left\{\begin{array}{c}2\\1\end{array}\right\} + \ldots + \binom{r}{r-1}\left\{\begin{array}{c}r-1\\1\end{array}\right\}\right)\frac{t^2}{2} + \ldots + \binom{r}{r-1}\left\{\begin{array}{c}r-1\\r-1\end{array}\right\}\frac{t^r}{r}.$$

Equating coefficients of t^{m+1} on both sides of the equation gives

$$\left\{{n+1 \atop m+1}\right\} = \frac{1}{m+1}\sum_{j=m}^{n}\binom{n+1}{j}\left\{{j \atop m}\right\}. \quad \square$$

Example 10.30. Let $n=6$ and $m=3$. Then $\left\{{n+1 \atop m+1}\right\} = \left\{{7 \atop 4}\right\} = 350$. Also, by Theorem 10.29, we have

$$\frac{1}{m+1}\sum_{j=m}^{n}\binom{n+1}{j}\left\{{j \atop m}\right\} = \frac{1}{4}\sum_{j=3}^{6}\binom{7}{j}\left\{{j \atop 3}\right\}$$

$$= \frac{1}{4}\left(\binom{7}{3}\left\{{3 \atop 3}\right\} + \binom{7}{4}\left\{{4 \atop 3}\right\} + \binom{7}{5}\left\{{5 \atop 3}\right\} + \binom{7}{6}\left\{{6 \atop 3}\right\}\right)$$

$$= \frac{1}{4}\left((35)(1) + (35)(6) + (21)(25) + (7)(90)\right)$$

$$= \frac{1}{4}(35 + 210 + 525 + 630)$$

$$= \frac{1}{4}(1400)$$

$$= 350.$$

Exercises

1. Show that $h_6(t) * h_6(t) = 0$.

2. Use the recurrence relation in Theorem 10.27 and Table 10.2 to calculate $f_8(t)$.

3. Calculate $\tau_3(36)$ using $\tau_0(n) = 1$ and $\tau_{r+1}(n) = \sum_{d|n}\tau_r(d)$.

4. Calculate $\tau_3(36)$ using Theorems 10.6 and 10.8.

5. Use Table 10.1 to verify Corollary 10.3 for $n=6$ and $m=4$: $\sum_{k=0}^{3}(-1)^k\binom{4}{k}\tau_{3-k}(6) = 0$.

Chapter 11. Generalized Exponential Bell Partition Polynomials

The exponential Bell partition polynomials $B_n = \sum \dfrac{n!}{k_1!(1!)^{k_1} k_2!(2!)^{k_2} \ldots k_n!(n!)^{k_n}} x_1^{k_1} x_2^{k_2} \ldots x_n^{k_n}$ are multivariable generating functions for counting the numbers of partitions of a finite set of n distinct elements into any number of subsets (Charalambides [2]). If, however, the elements, or objects, are not distinct, the analogous polynomials are naturally considered as *generalized* exponential Bell partition polynomials. Using the number theoretic paradigm, we can develop a logarithmic generating function for the generalized exponential Bell partition polynomials and a recurrence relation for calculating the polynomials. The classical exponential Bell partition polynomials are then obtained as a special case of a more general theorem (Theorem 11.6).

Since we have been developing the theory of enumerating object distributions using logarithmic generating functions, our approach will be to extend the definition of the object distribution generating function so that it accounts for occupancies within the cells or boxes. The natural metaphor that we have been using is to consider the placement of objects, such as colored balls, into identical boxes with no box empty. There is no restriction assumed on the objects. Any collection of objects, or colored balls, is allowed, and repetition of objects is also allowed. This is where we deviate from the traditional exponential Bell partition polynomials. Rather than requiring that the objects be distinct, repetitions of object types is allowed. So, our polynomials will necessarily be generalizations of the exponential Bell partition polynomials.

Definition 11.1. If $n = p_1^{\alpha_1} p_2^{\alpha_2} \ldots p_k^{\alpha_k}$, for distinct primes p_j, then we define $e(n) = \sum_j \alpha_j$. In other words, $e(n)$ counts the sum of the prime exponents in the unique prime factorization of n.

Example 11.2. Let $n = 45 = 3^2 \cdot 5^1$. Then $e(45) = 2 + 1 = 3$, the sum of the primes' exponents.

The enumeration logarithmic generating function for counting the number of ways to distribute objects of specification n into identical boxes with no box empty is given by:

$$G(x, t) = \sum_{n=1}^{\infty} g_n(t) x^{\log n} = \prod_{d>1} \frac{1}{(1-tx^{\log d})}. \quad (11.1)$$

We can now add an "object occupancy counter," $\lambda_{e(d)}$ to $G(x, t)$ to obtain:

$$G(x, \lambda, t) = \sum_{n=1}^{\infty} g_n(\lambda, t) x^{\log n} = \prod_{d>1} \frac{1}{(1-t\lambda_{e(d)} x^{\log d})}. \quad (11.2)$$

The parameter $\lambda_{e(d)}$ counts the number of objects in a box. For example, if we expand (11.2) and obtain a term like $2\lambda_1\lambda_2 t^2$, then this term tells us that we have two distributions of objects into two boxes (since we have t^2) such that one box has one object (λ_1) and one box has two objects (λ_2).

In (11.2), we represent the object counters as a vector, λ, because a general term in the expansion of (11.2) may contain, for example, λ_1, λ_2, λ_3, etc.

Example 11.3. Suppose we have two red balls and one blue ball. Let a red ball be represented by the prime number 2, and let a blue ball be represented by the prime number 3. A specification number for the multiset of colored balls $\{r, r, b\} = \{2, 2, 3\}$ is $n = 2 \cdot 2 \cdot 3 = 2^2 \cdot 3 = 12$. To find the number of ways to distribute these objects into identical boxes with no box empty, we expand the logarithmic generating function (11.2) and look at the term for $x^{\log 12}$. This term is

$$(\lambda_3 t + 2\lambda_1\lambda_2 t^2 + \lambda_1^3 t^3) x^{\log 12}. \quad (11.3)$$

Thus, $g_{12}(\lambda, t) = g_{12}((\lambda_1, \lambda_2, \lambda_3), t) = \lambda_3 t + 2\lambda_1\lambda_2 t^2 + \lambda_1^3 t^3$. So, for example, to find the number of distributions into *two* boxes, we look at the coefficient of t^2, which is $2\lambda_1\lambda_2$. This tells us that there are two distributions such that one box has one object (indicated by λ_1) and one box has two objects (indicated by λ_2). The two distributions are $(\{r\}, \{r, b\})$ and $(\{b\}, \{r, r\})$.

Example 11.4. Suppose we have four *distinct* objects to distribute into identical boxes with no box empty. We can represent the four objects by the prime numbers 2, 3, 5, and 7. This gives us a specification number $n = 2 \cdot 3 \cdot 5 \cdot 7 = 210$. To find the number of ways to distribute these four distinct objects into identical boxes with no box empty, we expand the logarithmic generating function (11.2) and read off the term for $x^{\log 210}$: $(\lambda_4 t + (4\lambda_1\lambda_3 + 3\lambda_2^2)t^2 + 6\lambda_1^2\lambda_2 t^3 + \lambda_1^4 t^4)x^{\log 210}$.

Thus, $g_{210}(\lambda, t) = \lambda_4 t + (4\lambda_1\lambda_3 + 3\lambda_2^2)t^2 + 6\lambda_1^2\lambda_2 t^3 + \lambda_1^4 t^4$. So, the number of ways to distribute the four distinct objects into k boxes is given by the coefficient of t^k. For *two* boxes, the coefficient of t^2 is $4\lambda_1\lambda_3 + 3\lambda_2^2$. This tells us that there are four distributions into two boxes such that one box contains one object (indicated by λ_1) and one box contains three objects (indicated by λ_3). Also, there are three distributions into two boxes such that one box contains two objects (indicated by one of the λ_2's) and the other box also contains two objects (indicated by the other λ_2) (note that $\lambda_2 \cdot \lambda_2 = \lambda_2^2$).

If we are only concerned with counting the total number of distributions into identical boxes, without regard to their occupancies, we clearly have $g_n(1, t) = g_n(t)$. In other words, set all the $\lambda_j = 1$. In this Example 11.4, we have $g_{210}(t) = g_{210}(1, t) = t + 7t^2 + 6t^3 + t^4$.

An astute observer might notice that when the objects are distinct, the coefficients of t^k in $g_n(\lambda, t)$ are the *exponential Bell partition polynomials* (Charalambides [2]). For example, we have

$$g_{210}(\lambda, t) = \lambda_4 t + (4\lambda_1\lambda_3 + 3\lambda_2^2)t^2 + 6\lambda_1^2\lambda_2 t^3 + \lambda_1^4 t^4$$
$$= B_{4,1}t + B_{4,2}t^2 + B_{4,3}t^3 + B_{4,4}t^4$$
$$= \sum_{k=1}^{4} B_{4,k}t^k.$$

Then, if we set each $\lambda_j = 1$, $g_{210}(\lambda, 1) = \sum_{k=1}^{4} B_{4,k} = B_4$. There is an explicit formula for the Bell number B_n, namely

$$B_n = \sum \frac{n!}{k_1!(1!)^{k_1} k_2!(2!)^{k_2} \ldots k_n!(n!)^{k_n}}, \tag{11.4}$$

where the summation is extended over all partitions of n, i.e., over all non-negative integer solutions $(k_1, k_2, \ldots, k_n)$ of the equation $k_1 + 2k_2 + \ldots + nk_n = n$. In summary, we have the following observation:

The polynomials $g_n(\lambda, t)$ in the logarithmic generating function given by (11.2) are generalizations of the exponential Bell partition polynomials. If the set of objects specified by n are distinct, then the polynomials $g_n(\lambda, t)$ are precisely the exponential Bell partition polynomials.

Our next task is to derive a recurrence relation for the *generalized* exponential Bell partition polynomials so that we can calculate them without having to expand the generating function (11.2).

Lemma 11.5. $t \cdot \dfrac{d}{dt} g_n(\lambda, t) = \displaystyle\sum_{\substack{d^k \mid n \\ d^k > 1}} t^k \lambda_{e(d)}^k g_{\frac{n}{d^k}}(\lambda, t)$

Proof. By (11.2), we have

$$\prod_{d>1} \frac{1}{(1 - t\lambda_{e(d)} x^{\log d})} = \sum_{n=1}^{\infty} g_n(\lambda, t) x^{\log n}$$

$$\log\left(\prod_{d>1} \frac{1}{(1 - t\lambda_{e(d)} x^{\log d})}\right) = \log\left(\sum_{n=1}^{\infty} g_n(\lambda, t) x^{\log n}\right)$$

$$\sum_{d=2}^{\infty} \log(1 - t\lambda_{e(d)} x^{\log d})^{-1} = \log\left(\sum_{n=1}^{\infty} g_n(\lambda, t) x^{\log n}\right)$$

$$-\sum_{d=2}^{\infty} \log(1 - t\lambda_{e(d)} x^{\log d}) = \log\left(\sum_{n=1}^{\infty} g_n(\lambda, t) x^{\log n}\right)$$

Taking the derivative of both sides with respect to t gives

$$\sum_{d=2}^{\infty} \frac{\lambda_{e(d)} x^{\log d}}{(1 - t\lambda_{e(d)} x^{\log d})} = \frac{\displaystyle\sum_{n=1}^{\infty} \frac{d}{dt} g_n(\lambda, t) x^{\log n}}{\displaystyle\sum_{n=1}^{\infty} g_n(\lambda, t) x^{\log n}}$$

Multiplying both sides by t gives

$$\sum_{d=2}^{\infty} \frac{t\lambda_{e(d)} x^{\log d}}{(1-t\lambda_{e(d)} x^{\log d})} = \frac{\sum_{n=1}^{\infty} t \frac{d}{dt} g_n(\lambda, t) x^{\log n}}{\sum_{n=1}^{\infty} g_n(\lambda, t) x^{\log n}}$$

Note that $\sum \frac{u_n}{1-u_n} = \sum (u_n + u_n^2 + u_n^3 + \ldots)$, so the previous equation becomes

$$\sum_{d=2}^{\infty} (t\lambda_{e(d)} x^{\log d} + t^2 \lambda_{e(d)}^2 x^{2\log d} + t^3 \lambda_{e(d)}^3 x^{3\log d} + \ldots) = \frac{\sum_{n=1}^{\infty} t \frac{d}{dt} g_n(\lambda, t) x^{\log n}}{\sum_{n=1}^{\infty} g_n(\lambda, t) x^{\log n}}$$

$$\left(\sum_{d=2}^{\infty} (t\lambda_{e(d)} x^{\log d} + t^2 \lambda_{e(d)}^2 x^{2\log d} + t^3 \lambda_{e(d)}^3 x^{3\log d} + \ldots) \right) \left(\sum_{n=1}^{\infty} g_n(\lambda, t) x^{\log n} \right) = \sum_{n=1}^{\infty} t \frac{d}{dt} g_n(\lambda, t) x^{\log n}$$

Now, if we multiply the left-hand side, and compare coefficients of $x^{\log n}$ on both sides of the last equation, we obtain the desired result: $t \cdot \frac{d}{dt} g_n(\lambda, t) = \sum_{\substack{d^k \mid n \\ d^k > 1}} t^k \lambda_{e(d)}^k g_{\frac{n}{d^k}}(\lambda, t)$ □

Note that when we use Lemma 11.5, summing over all $d^k \mid n$ (i.e., all d^k that divide n), we must sum over *all* d^k that divide n. For example, for $n = 4$, $d^k \in \{2^1, 2^2, 4^1\}$, where $d > 1$ and $k \geq 1$.

Now that we have Lemma 11.5, we can easily obtain a recurrence relation for the generalized exponential Bell partition polynomials $g_n(\lambda, t)$.

Theorem 11.6 (Generalized Exponential Bell Partition Polynomials). *A recurrence relation for the generalized exponential Bell partition polynomials, $g_n(\lambda, t)$, is given by*

$$g_n(\lambda, t) = \sum_{\substack{d^k \mid n \\ d^k > 1}} \lambda_{e(d)}^k \int_0^t u^{k-1} g_{\frac{n}{d^k}}(\lambda, u) \, du, \tag{11.5}$$

where $g_1(\lambda, t) = 1$, by definition.

Proof. Using Lemma 11.5, $t \cdot \dfrac{d}{dt} g_n(\lambda, t) = \sum\limits_{\substack{d^k | n \\ d^k > 1}} t^k \lambda_{e(d)}^k g_{\frac{n}{d^k}}(\lambda, t)$ divide both sides of by t (technically t cannot be zero, but t is being used as a formal variable in an algebraic ring), then integrate both sides of the resulting equation. When we integrate, the variable t in the integrand is replaced by a dummy variable of integration, u. Also, we use the fact that $g_n(\lambda, 0) = 0$ when we integrate from 0 to t. This gives the desired result. $\square$

Corollary 11.7. *For the special case where all the objects are distinct, Theorem 11.6 reduces to a recurrence relation for the exponential Bell partition polynomials. For $n = p_1 p_2 \ldots p_m$, for distinct primes p_j, we have*

$$B_m(\lambda, t) = g_n(\lambda, t) = \sum_{\substack{d|n \\ d>1}} \lambda_{e(d)} \int_0^t g_{\frac{n}{d}}(\lambda, u)\, du$$

$$B_m(\lambda) = B_m(\lambda_1, \lambda_2, \ldots, \lambda_m) = \sum_{\substack{d|n \\ d>1}} \lambda_{e(d)} \int_0^1 g_{\frac{n}{d}}(\lambda, u)\, du.$$

Proof. The result follows from Theorem 11.6. Since n is, by hypothesis, a product of distinct primes, any d^k that divides n, for $d^k > 1$, is also a product of distinct primes. Also, no divisor of n consists of a prime to a power greater than 1. So, all d^k that divide n have the form d^1, where $k = 1$.

Example 11.8. The exponential Bell partition polynomial $B_3(\lambda)$ is given by the expression $B_3(\lambda) = B_3(\lambda_1, \lambda_2, \lambda_3) = \lambda_3 + 3\lambda_1\lambda_2 + \lambda_1^3$. Let us calculate $B_3(\lambda)$ recursively using Corollary 11.7. For three distinct objects, let the object specification number be $n = p_1 p_2 p_3$ for distinct primes p_j. For example, let $n = 2 \cdot 3 \cdot 5 = 30$. The divisors of 30 are $d \in \{1, 2, 3, 5, 6, 10, 15, 30\}$. Suppose we also know that $B_1(\lambda_1, t) = \lambda_1 t = g_{p_1}(\lambda_1, t)$ and $B_2(\lambda_1, \lambda_2, t) = \lambda_2 t + \lambda_1^2 t^2 = g_{p_1 p_2}(\lambda_1, \lambda_2, t)$. Then, by Corollary 11.7, we have

$$B_3(\lambda) = \sum_{\substack{d|30 \\ d>1}} \lambda_{e(d)} \int_0^1 g_{\frac{30}{d}}(\lambda, u)\, du$$

$$= \binom{3}{1}\lambda_1 \int_0^1 g_{p_1 p_2}(\lambda, u)\, du + \binom{3}{2}\lambda_2 \int_0^1 g_{p_1}(\lambda, u)\, du + \binom{3}{3}\lambda_3 \int_0^1 g_1(\lambda, u)\, du$$

$$= \binom{3}{1}\lambda_1 \int_0^1 (\lambda_2 u + \lambda_1^2 u^2)\, du + \binom{3}{2}\lambda_2 \int_0^1 \lambda_1 u\, du + \binom{3}{3}\lambda_3 \int_0^1 1\, du$$

$$= \lambda_3 + 3\lambda_1 \lambda_2 + \lambda_1^3.$$

The binomial coefficients naturally arise because $g_{p_1 p_2 \ldots p_k}(\lambda, t)$ has the same value for any choice of k distinct primes. So, we end up choosing various combinations of primes from those that compose the specification number n. If, for example, $n = p_1 p_2 p_3$, then $g_{p_1 p_2} + g_{p_2 p_3} + g_{p_1 p_3}$ can be written as

$$\binom{3}{2} g_{p_1 p_2}.$$

Corollary 11.9. *For a set of distinct objects, the exponential Bell partition polynomials are given by*

$$B_m(\lambda) = \sum_{k=1}^{m} \binom{m}{k} \lambda_k \int_0^1 B_{m-k}(\lambda, u)\, du$$

where $B_0 = 1$.

Proof. Using Corollary 11.7, and as shown in Example 11.8, for $n = p_1 p_2 \ldots p_m$, we have

$$B_m(\lambda, t) = g_n(\lambda, t)$$

$$= \sum_{k=1}^{m} \binom{m}{k} \lambda_k \int_0^t g_{\frac{p_1 p_2 \ldots p_m}{p_1 p_2 \ldots p_k}}(\lambda, u)\, du.$$

And we have

$$B_m(\lambda) = \sum_{k=1}^{m} \binom{m}{k} \lambda_k \int_0^1 g_{\frac{p_1 p_2 \ldots p_m}{p_1 p_2 \ldots p_k}}(\lambda, u)\, du$$

$$= \sum_{k=1}^{m} \binom{m}{k} \lambda_k \int_0^1 g_{p_1 p_2 \ldots p_{m-k}}(\lambda, u)\, du$$

$$= \sum_{k=1}^{m} \binom{m}{k} \lambda_k \int_0^1 B_{m-k}(\lambda, u)\, du$$

where $B_0 = 1$. □

In the conventional theory of partition polynomials, the exponential Bell partition polynomials are given by $B_n = \sum_{k=1}^{n} B_{n,k}$. So, $B_n(\lambda, t) = \sum_{k=1}^{n} B_{n,k} t^k$. For example, we have

$$B_4(\lambda, t) = \sum_{k=1}^{4} B_{4,k} t^k$$
$$= B_{4,1} t + B_{4,2} t^2 + B_{4,3} t^3 + B_{4,4} t^4$$
$$= \lambda_4 t + (4\lambda_1 \lambda_3 + 3\lambda_2^2) t^2 + 6\lambda_1^2 \lambda_2 t^3 + \lambda_1^4 t^4.$$

What does this result mean? The 4 in $B_4(\lambda, t)$ means that we have four distinct objects. The coefficient of t^k gives the number of ways to distribute the four distinct objects into precisely k identical boxes with no box empty. Here, for two boxes, we look at the coefficient of t^2, which is $4\lambda_1 \lambda_3 + 3\lambda_2^2$. This tells us that there are 4 distributions of the objects into two boxes such that one box has one object (indicated by λ_1) and one box has three objects (indicated by λ_3). We see this from the term $4\lambda_1 \lambda_3$. Similarly, the term $3\lambda_2^2 = 3\lambda_2 \lambda_2$ tells us that there are three distributions such that one box has two objects (indicated by λ_2) and another box has two objects (also λ_2).

Note that the enumeration polynomials $g_n(\lambda, t)$, in Theorem 11.6, are more general than the Bell polynomials, $B_m(\lambda, t)$ in Corollary 11.7, because the objects used in distributions counted by $g_n(\lambda, t)$ need not be distinct. The objects used in distributions counted by $g_n(\lambda, t)$ can have repetitions, or multiplicities, among the object types.

Example 11.10. Let us use Corollary 11.9 to calculate the exponential Bell partition polynomial B_4. By Corollary 11.9, we have

$$B_4(\lambda) = \sum_{k=1}^{4} \binom{4}{k} \lambda_k \int_0^1 B_{4-k}(\lambda, u)\, du$$

$$= \binom{4}{1} \lambda_1 \int_0^1 B_3(\lambda, u)\, du + \binom{4}{2} \lambda_2 \int_0^1 B_2(\lambda, u)\, du + \binom{4}{3} \lambda_3 \int_0^1 B_1(\lambda, u)\, du + \binom{4}{4} \lambda_4 \int_0^1 B_0(\lambda, u)\, du$$

$$= \binom{4}{1} \lambda_1 \int_0^1 (\lambda_3 u + 3\lambda_1\lambda_2 u^2 + \lambda_1^3 u^3)\, du + \binom{4}{2} \lambda_2 \int_0^1 (\lambda_2 u + \lambda_1^2 u^2)\, du + \binom{4}{3} \lambda_3 \int_0^1 \lambda_1 u\, du + \binom{4}{4} \lambda_4 \int_0^1 1\, du$$

$$= \binom{4}{1} \lambda_1 \left(\frac{\lambda_3}{2} + 3\frac{\lambda_1\lambda_2}{3} + \frac{\lambda_1^3}{4} \right) + \binom{4}{2} \lambda_2 \left(\frac{\lambda_2}{2} + \frac{\lambda_1^2}{3} \right) + \binom{4}{3} \lambda_3 \left(\frac{\lambda_1}{2} \right) + \binom{4}{4} \lambda_4$$

$$= \lambda_4 + 4\lambda_1\lambda_3 + 3\lambda_2^2 + 6\lambda_1^2\lambda_2 + \lambda_1^4.$$

We can easily insert the "box counter" variable, t, to obtain $B_4(\lambda, t)$, namely $B_4(\lambda, t) = \lambda_4 t + 4\lambda_1\lambda_3 t^2 + 3\lambda_2^2 t^2 + 6\lambda_1^2\lambda_2 t^3 + \lambda_1^4 t^4 = \lambda_4 t + (4\lambda_1\lambda_3 + 3\lambda_2^2)t^2 + 6\lambda_1^2\lambda_2 t^3 + \lambda_1^4 t^4$. The values for $B_3(\lambda, u) = \lambda_3 u + 3\lambda_1\lambda_2 u^2 + \lambda_1^3 u^3$, $B_2(\lambda, u) = \lambda_2 u + \lambda_1^2 u^2$, $B_1(\lambda, u) = \lambda_1 u$, and $B_0(\lambda, u) = 1$ can either be calculated recursively using Corollary 11.9, or we can look up the polynomials in a reference table of Bell polynomials (e.g., [2], Table 11.1, p. 417).

Theorem 11.6 gives us a recurrence relation for calculating the generalized exponential Bell partition polynomials. We now will derive an alternative recurrence relation for calculating these polynomials. Unlike the previous recurrence relation (Theorem 11.6), this recurrence has the advantage that we only need to sum over the positive integer divisors of n. For the special case of *distinct* objects, Theorem 11.11 reduces to a well-known recurrence for the exponential Bell partition polynomials.

Theorem 11.11 (Recurrence Relation for the Generalized Exponential Bell Partition Polynomials). *A recurrence relation for the generalized exponential Bell partition polynomials, $g_n(\lambda, t)$, is given by*

$$g_n(\lambda, t) = \sum_{\substack{d \mid n \\ d \neq n}} \int_0^t \frac{1}{x} g_d(\lambda, x) P_{\frac{n}{d}}(\lambda, x)\, dx, \tag{11.6}$$

where, $g_1(\lambda, t) = 1$ by definition, and $P_m(\lambda, t) = \sum_{k=1}^{\infty} N(m, k) t^k \lambda_{e(m^{1/k})}^k$ where $N(m, k) = 1$ if the positive integer m can be represented as the k-th power of a positive integer, and $N(m, k) = 0$ otherwise. Note that the objects can have any specification. They can be identical, distinct, or of mixed specification. Theorem 11.11 is entirely general with regard to the object set specified by the object specification number n.

Proof. By (11.2), we have

$$\prod_{d \geq 1} \frac{1}{(1 - t\lambda_{e(d)} x^{\log d})} = \sum_{n=1}^{\infty} g_n(\lambda, t) x^{\log n}$$

$$\log\left(\prod_{d \geq 1} \frac{1}{(1 - t\lambda_{e(d)} x^{\log d})}\right) = \log\left(\sum_{n=1}^{\infty} g_n(\lambda, t) x^{\log n}\right)$$

$$\sum_{d=2}^{\infty} \log(1 - t\lambda_{e(d)} x^{\log d})^{-1} = \log\left(\sum_{n=1}^{\infty} g_n(\lambda, t) x^{\log n}\right)$$

$$-\sum_{d=2}^{\infty} \log(1 - t\lambda_{e(d)} x^{\log d}) = \log\left(\sum_{n=1}^{\infty} g_n(\lambda, t) x^{\log n}\right)$$

Taking the derivative of both sides with respect to t gives

$$\sum_{d=2}^{\infty} \frac{\lambda_{e(d)} x^{\log d}}{(1 - t\lambda_{e(d)} x^{\log d})} = \frac{\sum_{n=1}^{\infty} \frac{d}{dt} g_n(\lambda, t) x^{\log n}}{\sum_{n=1}^{\infty} g_n(\lambda, t) x^{\log n}}$$

$$\left(\sum_{n=1}^{\infty} g_n(\lambda, t) x^{\log n}\right)\left(\sum_{d=2}^{\infty} \frac{\lambda_{e(d)} x^{\log d}}{(1 - t\lambda_{e(d)} x^{\log d})}\right) = \sum_{n=1}^{\infty} \frac{d}{dt} g_n(\lambda, t) x^{\log n} \quad (11.7)$$

But what is $\sum_{d=2}^{\infty} \frac{\lambda_{e(d)} x^{\log d}}{(1 - t\lambda_{e(d)} x^{\log d})}$? First, note that multiplying by t gives

$t \cdot \sum_{d=2}^{\infty} \frac{\lambda_{e(d)} x^{\log d}}{(1 - t\lambda_{e(d)} x^{\log d})} = \sum_{d=2}^{\infty} (t\lambda_{e(d)} x^{\log d} + t^2 \lambda_{e(d)}^2 x^{2\log d} + t^3 \lambda_{e(d)}^3 x^{3\log d} + ...)$. When we expand the sum and collect like terms, we find that for each $x^{\log n}$ we will get a contribution to $x^{\log n}$ each time n has the form d^k for some positive integer d. The coefficient of $x^{\log 16}$, for example, is $t\lambda_4 + t^2\lambda_2^2 + t^4\lambda_1^4$ since $t\lambda_4 x^{\log 16} + t^2\lambda_2^2 x^{2\log 4} + t^4\lambda_1^4 x^{4\log 2} = (t\lambda_4 + t^2\lambda_2^2 + t^4\lambda_1^4)x^{\log 16}$. Thus, we have

$$\sum_{d=2}^{\infty} \frac{t\lambda_{e(d)} x^{\log d}}{(1 - t\lambda_{e(d)} x^{\log d})} = \sum_{n=2}^{\infty} \left(\sum_{k=1}^{\infty} N(n, k) t^k \lambda_{e(n^{1/k})} \right) x^{\log n}$$

$$= \sum_{n=2}^{\infty} (P_n(\lambda, t)) x^{\log n}$$

where we define $P_n(\lambda, t) = \sum_{k=1}^{\infty} N(n, k) t^k \lambda_{e(n^{1/k})}^k$. Continuing with the proof, multiplying equation (11.7) by t gives

$$\left(\sum_{n=1}^{\infty} g_n(\lambda, t) x^{\log n} \right) \left(\sum_{n=2}^{\infty} P_n(\lambda, t) x^{\log n} \right) = \sum_{n=1}^{\infty} t \frac{d}{dt} g_n(\lambda, t) x^{\log n}.$$

By the convolution property for the logarithmic generating function, this last equation implies

$$t \frac{d}{dt} g_n(\lambda, t) = g_n(\lambda, t) * P_n(\lambda, t) \tag{11.8}$$

for $n > 1$. Note that we must take $n > 1$ because the $P_n(\lambda, t) x^{\log n}$ are summed from $n = 2$ to infinity. Thus, equation (11.8) gives us

$$t \frac{d}{dt} g_n(\lambda, t) = \sum_{\substack{d \mid n \\ d \neq n}} g_d(\lambda, t) P_{\frac{n}{d}}(\lambda, t). \tag{11.9}$$

Note that $d \neq n$ because if $d = n$ we would get a term $P_1(\lambda, t)$, but (11.8) requires $n > 1$ in $P_n(\lambda, t)$. If we now divide both sides of (11.9) by the formal variable t (where $t \neq 0$) we get

$$\frac{d}{dt} g_n(\lambda, t) = \frac{1}{t} \sum_{\substack{d \mid n \\ d \neq n}} g_d(\lambda, t) P_{\frac{n}{d}}(\lambda, t).$$

Finally, integrating both sides of this last equation with respect to a dummy variable x, we get the final result:

$$g_n(\lambda, t) = \sum_{\substack{d \mid n \\ d \neq n}} \int_0^t \frac{1}{x} g_d(\lambda, x) P_{\frac{n}{d}}(\lambda, x) \, dx \quad \square$$

Example 11.12. Suppose we have four *identical* objects, such as four red balls, and we want to count the number of ways to distribute the objects into identical boxes with no box empty. The object set can be represented by the multiset $S = \{r, r, r, r\}$. If we let the prime number 2 represent a red ball, then the object multiset becomes $S = \{2, 2, 2, 2\}$. A specification number for this set of objects is $n = 2 \cdot 2 \cdot 2 \cdot 2 = 16$. We want to calculate the object distribution polynomial $g_{16}(\lambda, t)$. The divisors, d, of 16 are $\{1, 2, 4, 8, 16\}$. By Theorem 11.11, we have

$$g_{16}(\lambda, t) = \sum_{\substack{d|16 \\ d \neq 16}} \int_0^t \frac{1}{x} g_d(\lambda, x) P_{\frac{16}{d}}(\lambda, x)\, dx$$

$$= \int_0^t \frac{1}{x} g_1(\lambda, x) P_{16}(\lambda, x)\, dx + \int_0^t \frac{1}{x} g_2(\lambda, x) P_8(\lambda, x)\, dx + \qquad (11.10)$$

$$\int_0^t \frac{1}{x} g_4(\lambda, x) P_4(\lambda, x)\, dx + \int_0^t \frac{1}{x} g_8(\lambda, x) P_2(\lambda, x)\, dx$$

Now we need to calculate $P_2(\lambda, x)$, $P_4(\lambda, x)$, $P_8(\lambda, x)$, and $P_{16}(\lambda, x)$ using

$$P_m(\lambda, t) = \sum_{k=1}^{\infty} N(m, k) t^k \lambda_{e(m^{1/k})}^k .$$

- Since 2 can be represented as a positive integer power in only one way, namely $2 = 2^1$, we have
  $P_2(\lambda, x) = N(2, 1) x^1 \lambda_{e(2^{1/1})}^1 = 1 \cdot x \lambda_{e(2)} = x\lambda_1$.

- Since 4 can be represented as a positive integer power in two ways, namely $4 = 4^1$ and $4 = 2^2$, we have $P_4(\lambda, x) = N(4, 1) x^1 \lambda_{e(4^{1/1})}^1 + N(4, 2) x^2 \lambda_{e(4^{1/2})}^2 = x\lambda_{e(4)} + x^2 \lambda_{e(2)}^2 = x\lambda_2 + x^2 \lambda_1^2$.

- Since 8 can be represented as a positive integer power in two ways, namely $8 = 8^1$ and $8 = 2^3$, we have $P_8(\lambda, x) = N(8, 1) x^1 \lambda_{e(8^{1/1})}^1 + N(8, 3) x^3 \lambda_{e(8^{1/3})}^3 = x\lambda_{e(8)} + x^3 \lambda_{e(2)}^3 = x\lambda_3 + x^3 \lambda_1^3$.

- Since 16 can be represented as a positive integer power in three ways, namely $16 = 16^1$, $16 = 4^2$, and $16 = 2^4$, we have $P_{16}(\lambda, x) = N(16, 1) x^1 \lambda_{e(16^{1/1})}^1 + N(16, 2) x^2 \lambda_{e(16^{1/2})}^2 + N(16, 4) x^4 \lambda_{e(16^{1/4})}^4 =$
  $x\lambda_{e(16)} + x^2 \lambda_{e(4)}^2 + x^4 \lambda_{e(2)}^4 = x\lambda_4 + x^2 \lambda_2^2 + x^4 \lambda_1^4$.

Substituting these P expressions into (11.10), we obtain

$$g_{16}(\lambda, t) = \int_0^t \frac{1}{x} g_1(\lambda, x)(x\lambda_4 + x^2\lambda_2^2 + x^4\lambda_1^4)\,dx + \int_0^t \frac{1}{x} g_2(\lambda, x)(x\lambda_3 + x^3\lambda_1^3)\,dx +$$
$$\int_0^t \frac{1}{x} g_4(\lambda, x)(x\lambda_2 + x^2\lambda_1^2)\,dx + \int_0^t \frac{1}{x} g_8(\lambda, x)(x\lambda_1)\,dx \qquad (11.11)$$

We now use the presumably known object distribution polynomials (these can also be calculated using Theorem 11.11):

- $g_1(\lambda, x) = 1$
- $g_2(\lambda, x) = x\lambda_1$
- $g_4(\lambda, x) = x\lambda_2 + x^2\lambda_1^2$
- $g_8(\lambda, x) = x\lambda_3 + x^2\lambda_1\lambda_2 + x^3\lambda_1^3$

Substituting these g expressions into (11.11) and performing the integrations gives, after simplification, the final result:

$$g_{16}(\lambda, t) = \lambda_4 t + (\lambda_2^2 + \lambda_1\lambda_3)t^2 + \lambda_1^2\lambda_2 t^3 + \lambda_1^4 t^4. \qquad (11.12)$$

In particular, we have $g_{16}(\lambda, 1) = g_{2^4}(\lambda, 1) = \lambda_4 + \lambda_2\lambda_2 + \lambda_1\lambda_3 + \lambda_1\lambda_1\lambda_2 + \lambda_1\lambda_1\lambda_1\lambda_1$, and these are the *integer partitions of 4*: 4, 2 + 2, 1 + 3, 1 + 1 + 2, and 1 + 1 + 1 + 1. In other words, the number of ways to distribute four identical objects into identical boxes, with no box empty, is the same as the number of ways to partition the number 4 as a sum of positive integers.

Example 11.13. Suppose we have two red balls and one blue ball. The object set can be represented by the multiset $S = \{r, r, b\}$ or $S = \{2, 2, 3\}$ using prime numbers to represent each object type. A specification number for this set of objects is $n = 2 \cdot 2 \cdot 3 = 12$. We want to calculate the object distribution polynomial $g_{12}(\lambda, t)$, which counts the number of ways to distribute these objects into identical boxes with no box empty. If we use Theorem 11.11, following the same approach as illustrated in Example 11.12, we will obtain the object distribution polynomial $g_{12}(\lambda, t) = \lambda_3 t + 2\lambda_1\lambda_2 t^2 + \lambda_1^3 t^3$. This tells us that there is one distribution into one box containing all three objects (indicated by the term $\lambda_3 t$). There are two distributions of the objects into precisely two boxes such that one box has one object and the other box has two objects (indicated by the term $2\lambda_1\lambda_2 t^2$). There is one distribution of the objects into precisely three boxes such that each box contains one object (indicated by the term $\lambda_1^3 t^3$). The distributions look like this:

$$\{r, r, b\}, \{\{r\},\{r, b\}\}, \{\{b\}, \{r, r\}\}, \{\{r\}, \{r\}, \{b\}\}.$$

The order of the objects within each box, or cell, does not matter.

Exercises

1. Use Corollary 11.9 to inductively build the polynomial $B_5(\lambda)$. Note that $B_5(\lambda) = B_5(\lambda, t)$ for $t = 1$.

2. Explain what each term means in the polynomial $B_3(\lambda) = \lambda_3 + 3\lambda_1\lambda_2 + \lambda_1^3$. What does λ_j represent?

3. What does $g_n(\lambda, t)$ become if the specification nmber n represents k distinct objects?

4. What special combinatorial numbers do we get if we let $\lambda_j = 1$ for all j in the polynomials $B_m(\lambda_1, \lambda_2, \ldots, \lambda_m)$?

5. Use Theorem 11.11 to calculate $g_{18}(\lambda, t)$, where, since $18 = 2 \cdot 3^2$, we have, say, one red ball and two blue balls: $S = \{r, b, b\} = \{2, 3, 3\}$.

Chapter 12. Distributions in Identical Boxes with Ordered Occupancies

In previous chapters, we studied the problem of enumerating the ways to distribute objects of arbitrary specification, or types, into identical boxes. The order of the objects within each box, or cell, was considered unimportant. In this chapter, we study the enumeration problem where the order of the objects within each cell is considered important. Basically, the objects within a box can be permuted, so this problem imposes an additional occupancy constraint on the objects within the boxes. To solve this problem, we will modify the logarithmic generating function that we have been using to study all of these various distribution and occupancy problems using a number theoretic approach. The set of objects is encoded in a specification number, n, and we want to find the corresponding object distribution polynomials, $f_n(t)$, that count the number of possible distributions subject to the stated constraints. As usual, if $f_n(t)$ is one of the object distribution polynomials, then the coefficient of t^k is the number of ways to distribute the objects into precisely k boxes with no box empty. Although an explicit formula for $f(n)$, which is just $f_n(1)$, appears to be difficult to obtain, we will derive a general recurrence relation that will allow us to calculate the polynomials $f_n(t)$. We will then examine special cases of the general problem.

Suppose that we have a set of objects, such as colored balls, and we want to count the number of ways to distribute the objects into identical, or indistinguishable, boxes, with no box empty, such that the order of the objects within each box is important. Given a set, or specification, of objects, how can we count the number of such distributions? Consider, for example, the case where we have three objects, say two red balls and one blue ball. The balls (objects) of a given color (type) are identical to each other. In other words, all red balls are the same, all blue balls are the same, and so on. If we distribute these two red balls and one blue ball into identical boxes, with no box empty, we can use trial and error to count the number of possible distributions. This example is easy, because we only have three objects. There are seven possible distributions: three distributions of the objects into one box, three distributions of the objects into two boxes, and one distribution of the objects into three boxes. Remember, we are considering the order of the objects in the boxes to be significant, so we have to count all the possible permutations of the objects within their respective boxes. The seven distributions for this problem are (1) $\{r, r, b\}$, (2) $\{r, b, r\}$, (3) $\{b, r, r\}$, (4) $\{r\}\{r, b\}$, (5) $\{r\}\{b, r\}$, (6) $\{b\}\{r, r\}$, and (7) $\{r\}\{r\}\{b\}$. Note that the order of the objects within each box is important; however, since the boxes themselves are identical, the order of the boxes does not matter. That is the problem that we are currently studying. We will approach this general enumeration problem using the logarithmic generating function that we have been using to count other kinds of distributions of objects into boxes.

In previous chapters we used the logarithmic generating function to calculate object distribution polynomials, $g_n(t)$, that count the number of ways to distribute objects, of specification n, into identical boxes with no box empty:

$$G(x, t) = \sum_{n=1}^{\infty} g_n(t) x^{\log n} = \prod_{d>1} \frac{1}{(1 - tx^{\log d})}. \tag{12.1}$$

The number n is called a *specification number* which encodes the set of objects. How can we modify this generating function to calculate the object distribution polynomials $f_n(t)$, where the order of the objects within boxes matters?

When we expand (12.1), a typical term looks like $mt^k x^{\log d}$. This term tells us that there are precisely m ways to distribute objects of specification d into k identical boxes with no box empty. Let's look at a simple example. Suppose we have two red balls and one blue ball. The multiset of objects, or colored balls, is $S = \{r, r, b\}$, where r = red and b = blue. Next, we assign a unique prime number to each object type, or color, say $r = 2$ and $b = 3$. The specification number for these objects is defined to be $d = r \cdot r \cdot b = 2 \cdot 2 \cdot 3 = 12$. Since there is only one way to put all of these objects into $k = 1$ box, the corresponding term in the enumeration generating function is $mt^k x^{\log d} = 1 t^1 x^{\log 12}$, or just $tx^{\log 12}$. This term has coefficient 1 because there is only one way to distribute the objects into one box where the order of the objects within the box does not matter. But how do we modify the term if the order of the objects *does* matter within the box? Since, in this example, the objects are of type r, r, b, there are $\binom{3}{2, 1} = \frac{3!}{2!1!} = 3$ linear permutations of the objects within the box. So, we must modify the term $tx^{\log 12}$ by multiplying it by $\binom{3}{2, 1} = p(12)$. Here, the notation $p(12)$ means the number of permutations of the prime factors of 12, or $p(12) = p(\{2, 2, 3\}) = \binom{3}{2, 1}$, since there are $\binom{3}{2, 1}$ permutations of 2, 2, and 3.

Continuing with this example, let's build a generating function to count all of the ways to distribute objects of specification $S = \{r, r, b\}$ into identical boxes such that the order of the objects within each box does matter. Let $f_n(t)$ be the required object placement enumeration polynomial. Since the object set is $S = \{r, r, b\} = \{2, 2, 3\}$, the specification number is $n = 2 \cdot 2 \cdot 3 = 12$. So, we want to find the polynomial $f_{12}(t)$. (We are using a number-theoretic approach to study a combinatorics problem.) The non-unit divisors of 12 are $\{2, 3, 4, 6, 12\}$. The desired generating function is, by brute-force inspection, given by:

$(1 + p(2)tx^{\log 2} + p^2(2)t^2 x^{2\log 2})(1 + p(3)tx^{\log 3})(1 + p(4)tx^{\log 4})(1 + p(6)tx^{\log 6})(1 + p(12)tx^{\log 12}) =$

$(1 + tx^{\log 2} + t^2 x^{2\log 2})(1 + tx^{\log 3})(1 + tx^{\log 4})(1 + 2tx^{\log 6})(1 + 3tx^{\log 12})$.

If we multiply and collect like terms, we can read off the coefficient of $x^{\log 12}$, which will be $f_{12}(t)$. We get $f_{12}(t) = 3t + 3t^2 + t^3$. The coefficient of t^k gives the number of ordered distributions of the objects, having specification $n = 12$ (or $\{2, 2, 3\}$ or $\{r, r, b\}$) into precisely k identical boxes with no box empty. The total number of ordered distributions is $f_{12}(1) = 3 + 3 + 1 = 7$. This agrees with our previous example.

Definition 12.1 (Permutation Function). For positive integer $n > 1$, let $p(n)$ denote the number of linear permutations of the prime factors of n.

Example 12.2. Let $n = 180$. The unique prime factorization of n is $180 = 2^2 \cdot 3^2 \cdot 5$. Then $p(180)$ is the number of permutations of the elements in the multiset $S = \{2, 2, 3, 3, 5\}$. Note that S contains five elements, of which there are two of one kind, two of another kind, and one of a third kind. So,

$p(180) = \binom{5}{2, 2, 1} = \frac{5!}{2!2!1!} = 30$.

Theorem 12.3. *Let $f_n(t)$ be the object distribution polynomial that counts the number of ways to distribute objects of specification n into identical boxes, with no box empty, such that the order of the objects within each box matters. A generating function for $f_n(t)$ is given by*

$$F(x, t) = \sum_{n=1}^{\infty} f_n(t) x^{\log n} = \prod_{d>1} \frac{1}{(1 - p(d)t x^{\log d})}. \quad (12.2)$$

Proof. As discussed in the previous paragraphs, we modify the logarithmic generating function (12.1) by multiplying each term $tx^{\log d}$ by the permutation function $p(d)$ to permute the objects within each box. □

Definition 12.4 (Permutation Integer Power Polynomial). Let us define the *permutation integer power polynomial* for a positive integer $n > 1$ by

$$P_n(p, t) = \sum_{k=1}^{\infty} N(n, k) t^k p^k (n^{1/k}), \quad (12.3)$$

where $N(n, k) = 1$ if the positive integer n can be represented as the kth power of a positive integer ($k > 0$), and $N(n, k) = 0$ otherwise. The function $p(m)$, for positive integer $m > 1$, is the permutation function given in Definition 12.1.

Theorem 12.5. *Let $f_n(t)$ be the object distribution polynomial that counts the number of distributions of objects of specification n, with ordered occupancy, into identical boxes, or cells, with no box empty. A recurrence relation for $f_n(t)$ is given by*

$$f_n(t) = \sum_{\substack{d \mid n \\ d \neq n}} \int_0^t \frac{1}{x} f_d(x) P_{\frac{n}{d}}(p, x) \, dx, \quad (12.4)$$

where $P_n(p, t)$ is the permutation integer power polynomial and $f_1(t) = 1$, by definition.

Proof. By (12.2), we have

$$\prod_{d>1} \frac{1}{(1-tp(d)x^{\log d})} = \sum_{n=1}^{\infty} f_n(t)x^{\log n}$$

$$\log\left(\prod_{d>1} \frac{1}{(1-tp(d)x^{\log d})}\right) = \log\left(\sum_{n=1}^{\infty} f_n(t)x^{\log n}\right)$$

$$\sum_{d=2}^{\infty} \log(1-tp(d)x^{\log d})^{-1} = \log\left(\sum_{n=1}^{\infty} f_n(t)x^{\log n}\right)$$

$$-\sum_{d=2}^{\infty} \log(1-tp(d)x^{\log d}) = \log\left(\sum_{n=1}^{\infty} f_n(t)x^{\log n}\right)$$

Taking the derivative of both sides with respect to t gives

$$\sum_{d=2}^{\infty} \frac{p(d)x^{\log d}}{(1-tp(d)x^{\log d})} = \frac{\sum_{n=1}^{\infty} \frac{d}{dt}f_n(t)x^{\log n}}{\sum_{n=1}^{\infty} f_n(t)x^{\log n}}$$

$$\left(\sum_{n=1}^{\infty} f_n(t)x^{\log n}\right)\left(\sum_{d=2}^{\infty} \frac{p(d)x^{\log d}}{(1-tp(d)x^{\log d})}\right) = \sum_{n=1}^{\infty} \frac{d}{dt}f_n(t)x^{\log n} \qquad (12.5)$$

But what is $\sum_{d=2}^{\infty} \frac{p(d)x^{\log d}}{(1-tp(d)x^{\log d})}$? First, note that multiplying by t gives

$t \cdot \sum_{d=2}^{\infty} \frac{p(d)x^{\log d}}{(1-tp(d)x^{\log d})} = \sum_{d=2}^{\infty} (tp(d)x^{\log d} + t^2 p^2(d)x^{2\log d} + t^3 p^3(d)x^{3\log d} + ...)$. When we expand the sum and collect like terms, we find that for each $x^{\log n}$ we will get a contribution to $x^{\log n}$ each time n has the form d^k for some positive integer d. The coefficient of $x^{\log 16}$, for example, is
$tp(16) + t^2 p^2(4) + t^4 p^4(2)$ since $tp(16)x^{\log 16} + t^2 p^2(4)x^{2\log 4} + t^4 p^4(2)x^{4\log 2} =$
$(tp(16) + t^2 p^2(4) + t^4 p^4(2))x^{\log 16}$. Thus, we have

$$\sum_{d=2}^{\infty} \frac{tp(d)x^{\log d}}{(1-tp(d)x^{\log d})} = \sum_{n=2}^{\infty}\left(\sum_{k=1}^{\infty} N(n,k)t^k p^k(n^{1/k})\right)x^{\log n}$$

$$= \sum_{n=2}^{\infty} (P_n(p,t))x^{\log n}$$

where we define $P_n(p,t) = \sum_{k=1}^{\infty} N(n,k)t^k p^k(n^{1/k})$. Continuing with the proof, multiplying equation (12.5) by t gives

$$\left(\sum_{n=1}^{\infty} f_n(t) x^{\log n}\right)\left(\sum_{n=2}^{\infty} P_n(p, t) x^{\log n}\right) = \sum_{n=1}^{\infty} t \frac{d}{dt} f_n(t) x^{\log n}.$$

By the convolution property for the logarithmic generating function, this last equation implies

$$t \frac{d}{dt} f_n(t) = f_n(t) * P_n(p, t) \tag{12.6}$$

for $n > 1$. Note that we must take $n > 1$ because the $P_n(p, t) x^{\log n}$ are summed from $n = 2$ to infinity. Thus, equation (12.6) gives us

$$t \frac{d}{dt} f_n(t) = \sum_{\substack{d \mid n \\ d \neq n}} f_d(t) P_{\frac{n}{d}}(p, t). \tag{12.7}$$

Note that $d \neq n$ because if $d = n$ we would get a term $P_1(p, t)$, but (12.3) requires $n > 1$ in $P_n(p, t)$. If we now divide both sides of (12.7) by the formal variable t (where $t \neq 0$) we get

$$\frac{d}{dt} f_n(t) = \frac{1}{t} \sum_{\substack{d \mid n \\ d \neq n}} f_d(t) P_{\frac{n}{d}}(p, t).$$

Finally, integrating both sides of this last equation with respect to a dummy variable x, we get the final result:

$$f_n(t) = \sum_{\substack{d \mid n \\ d \neq n}} \int_0^t \frac{1}{x} f_d(x) P_{\frac{n}{d}}(p, x) \, dx \quad \square$$

Example 12.6. Suppose we have two red balls and one blue ball. In how many ways can we distribute these objects into identical boxes, with no box empty, such that the order of the objects within each box matters? Before we calculate the answer to this problem using Theorem 12.5, let's first observe what the solution looks like:

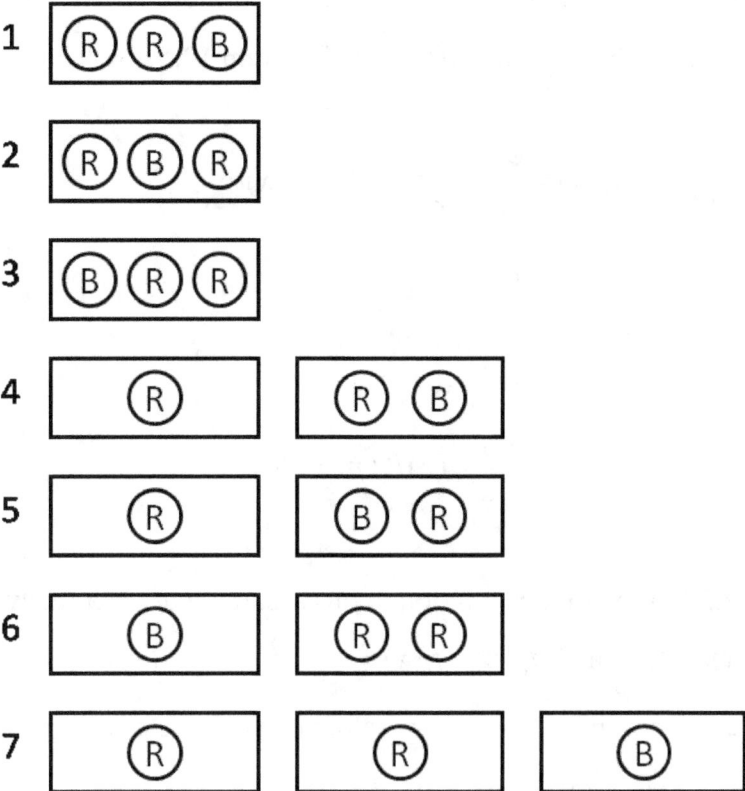

Figure 12.1. Ordered occupancy in identical boxes.

The order of the boxes does not matter, since the boxes are identical, but the order of the objects within each box does matter. There are clearly seven ways to distribute the objects with *ordered* occupancy in the identical boxes. Knowing this, we can immediately construct the object distribution (enumeration) polynomial, $f_n(t)$. The coefficient of t^k is the number of distributions into precisely k boxes. So, we have

$$f_n(t) = 3t + 3t^2 + t^3. \tag{12.8}$$

We solved this problem by brute-force construction of all the possible distributions. But how do we *calculate* the solution, $f_n(t)$?

Let the set of objects be $S = \{r, r, b\}$. If we represent a red ball by the prime number 2, and represent a blue ball by the prime number 3, then the set of objects is $S = \{r, r, b\} = \{2, 2, 3\}$. Then a specification number for the set of objects is $n = 2 \cdot 2 \cdot 3 = 12$. So, we want to calculate the object placement, or enumeration, polynomial $f_{12}(t)$. By Theorem 12.5, we have

$$\begin{aligned} f_{12}(t) &= \sum_{\substack{d|12 \\ d \neq 12}} \int_0^t \frac{1}{x} f_d(x) P_{\frac{12}{d}}(p, x) dx \\ &= \int_0^t \frac{1}{x} f_1(x) P_{12}(p, x) dx + \int_0^t \frac{1}{x} f_2(x) P_6(p, x) dx \\ &\quad + \int_0^t \frac{1}{x} f_3(x) P_4(p, x) dx + \int_0^t \frac{1}{x} f_4(x) P_3(p, x) dx \\ &\quad + \int_0^t \frac{1}{x} f_6(x) P_2(p, x) dx. \end{aligned} \quad (12.9)$$

Since Theorem 12.5 is a recurrence relation, let us assume that we already know the polynomials for $f_1(x)$, $f_2(x)$, $f_3(x)$, $f_4(x)$, and $f_6(x)$: $f_1(x) = 1$, $f_2(x) = x$, $f_3(x) = x$, $f_4(x) = x + x^2$, and $f_6(x) = 2x + x^2$. The polynomials $P_n(p, t)$ are calculated using Definition 12.4:

$$P_2(p, x) = \sum_{k \geq 1} N(2, k) x^k p^k (2^{1/k}) = 1 x p(2) = x$$

$$P_3(p, x) = \sum_{k \geq 1} N(3, k) x^k p^k (3^{1/k}) = 1 x p(3) = x$$

$$P_4(p, x) = \sum_{k \geq 1} N(4, k) x^k p^k (4^{1/k}) = N(4,1) x p(4) + N(4,2) x^2 p^2 (4^{1/2}) = x p(4) + x^2 p^2(2) = x + x^2$$

$$P_6(p, x) = \sum_{k \geq 1} N(6, k) x^k p^k (6^{1/k}) = x p(6) = 2x \quad \text{(Note that } p(6) = p(2 \cdot 3) = 2! = 2\text{)}$$

$$P_{12}(p, x) = \sum_{k \geq 1} N(12, k) x^k p^k (12^{1/k}) = x p(12) = 3x \quad \text{(Note that } p(12) = p(2 \cdot 2 \cdot 3) = \frac{3!}{2!1!} = 3\text{)}$$

Using these results, equation (12.9) becomes

$$\begin{aligned} f_{12}(t) &= \int_0^t \left(\tfrac{1}{x} 3x + \tfrac{1}{x} x \cdot 2x + \tfrac{1}{x} x(x + x^2) + \tfrac{1}{x}(x + x^2)x + \tfrac{1}{x}(2x + x^2)x \right) dx \\ &= \int_0^t \left(3 + 6x + 3x^2 \right) dx \\ &= 3t + 3t^2 + t^3. \end{aligned}$$

Note that this result agrees with equation (12.8) which was obtained by brute-force enumeration. □

Lemma 12.7. *Let $\varphi(r, t)$ be the object distribution polynomial that counts the number of ways to distribute r distinct objects into identical boxes such that the order of the objects within each box matters. A recurrence relation for $\varphi(r, t)$ is given by*

$$\varphi(r, t) = \sum_{k=0}^{r-1} \frac{r!}{k!} \int_0^t \varphi(k, x)\, dx, \qquad (12.10)$$

where $r \geq 1$ and $\varphi(0, t) = 1$.

Proof. If the objects are distinct, then the object specification number has the form $n = p_1 p_2 \ldots p_r$ for distinct primes p_j. Then, by Definition 12.4, we have

$$\begin{aligned}
P_n(p, t) &= \sum_{k=1}^{\infty} N(n, k) t^k p^k (n^{1/k}) \\
&= N(n, 1) t^1 p^1 (n^{1/1}) \\
&= t \cdot p(n) \\
&= t \cdot p(p_1 p_2 \ldots p_r) \\
&= t \cdot r!
\end{aligned}$$

Note that since a product of distinct primes can only be represented as a first power, we have $N(n, k) = N(n, 1) = 1$. So, we have

$$P_n(p, t) = t \cdot r! \quad \text{for } n = p_1 p_2 \ldots p_r \qquad (12.11)$$

By Theorem 12.5, we have

$$\begin{aligned}
f_n(t) &= \sum_{\substack{d \mid n \\ d \neq n}} \int_0^t \frac{1}{x} f_d(x) P_{\frac{n}{d}}(p, x)\, dx \\
&= \binom{r}{0} \int_0^t \frac{1}{x} f_1(x) P_{p_1 p_2 \ldots p_r}(p, x)\, dx + \binom{r}{1} \int_0^t \frac{1}{x} f_{p_1}(x) P_{p_2 \ldots p_r}(p, x)\, dx \\
&\quad + \binom{r}{2} \int_0^t \frac{1}{x} f_{p_1 p_2}(x) P_{p_3 p_4 \ldots p_r}(p, x)\, dx + \ldots + \binom{r}{r-1} \int_0^t \frac{1}{x} f_{p_1 p_2 \ldots p_{r-1}}(x) P_{p_r}(p, x)\, dx \\
&= \sum_{k=0}^{r-1} \binom{r}{k} \int_0^t \frac{1}{x} f_{p_1 p_2 \ldots p_k}(x) P_{p_{k+1} \ldots p_r}(p, x)\, dx
\end{aligned} \qquad (12.12)$$

The reason we have binomial coefficients is because of the symmetry in the sums. Note, for example, that when we have *distinct* objects, $f_{p_1 p_2}(x) = f_{p_2 p_3}(x) = \ldots$. These terms will occur in $\binom{r}{2}$ ways. Next, for simplicity, let us define $f_{p_1 p_2 \ldots p_j}(x) = \varphi(j, x)$. Combining this with the result in equation (12.11), equation (12.12) becomes

$$f_n(t) = \sum_{k=0}^{r-1} \binom{r}{k} \int_0^t \frac{1}{x} \varphi(k, x) \cdot x \cdot (r-k)! \, dx$$

$$= \sum_{k=0}^{r-1} \binom{r}{k} \int_0^t \varphi(k, x)(r-k)! \, dx$$

$$= \sum_{k=0}^{r-1} \frac{r!}{k!(r-k)!} \int_0^t \varphi(k, x)(r-k)! \, dx$$

$$= \sum_{k=0}^{r-1} \frac{r!}{k!} \int_0^t \varphi(k, x) \, dx$$

for $r \geq 1$. We define $\varphi(0, t) = 1$ to ensure that the recurrence works for trivial cases. □

Theorem 12.8 (Ordered Distributions of Distinct Objects in Identical Boxes). *Let $\varphi(r, t)$ be the object distribution polynomial that counts the number of ways to distribute r distinct objects into identical boxes, or cells, such that the order of the objects within each box matters. For $r > 1$ and $\varphi(1, t) = t$, a recurrence relation for $\varphi(r, t)$ is given by*

$$\varphi(r, t) = r \cdot \varphi(r-1, t) + r \cdot \int_0^t \varphi(r-1, x) \, dx \, . \tag{12.13}$$

Proof. For $r > 1$, Lemma 12.7 gives

$$\varphi(r, t) = \sum_{k=0}^{r-1} \frac{r!}{k!} \int_0^t \varphi(k, x) \, dx$$

$$= \sum_{k=0}^{r-2} \frac{r!}{k!} \int_0^t \varphi(k, x) \, dx + \frac{r!}{(r-1)!} \int_0^t \varphi(r-1, x) \, dx$$

$$= r \cdot \sum_{k=0}^{r-2} \frac{(r-1)!}{k!} \int_0^t \varphi(k, x) \, dx + r \cdot \int_0^t \varphi(r-1, x) \, dx$$

$$= r \cdot \varphi(r-1, t) + r \cdot \int_0^t \varphi(r-1, x) \, dx. \quad \square$$

The recurrence relation in Theorem 12.8, equation (12.13), together with the initial condition $\varphi(1, t) = t$, allows us to easily calculate a table of the object distribution polynomials $\varphi(r, t)$. See Table 12.1.

Table 12.1. Object Distribution polynomials for Ordered Occupancy in Identical Boxes

$\varphi(0, t) = 1$ by definition
$\varphi(1, t) = t$
$\varphi(2, t) = 2t + t^2$
$\varphi(3, t) = 6t + 6t^2 + t^3$
$\varphi(4, t) = 24t + 36t^2 + 12t^3 + t^4$
$\varphi(5, t) = 120t + 240t^2 + 120t^3 + 20t^4 + t^5$
$\varphi(6, t) = 720t + 1800t^2 + 1200t^3 + 300t^4 + 30t^5 + t^6$
$\varphi(7, t) = 5040t + 15120t^2 + 12600t^3 + 4200t^4 + 630t^5 + 42t^6 + t^7$
$\varphi(8, t) = 40320t + 141120t^2 + 141120t^3 + 58800t^4 + 11760t^5 + 1176t^6 + 56t^7 + t^8$
$\varphi(9, t) = 362880t + 1451520t^2 + 1693440t^3 + 846720t^4 + 211680t^5 + 28224t^6 + 2016t^7 + 72t^8 + t^9$

Some of the numbers in Table 12.1 are self-evident. For example, the coefficient of t in $\varphi(n, t)$ is $n!$, since there are $n!$ linear permutations of n distinct objects placed into one box. At the other extreme, the coefficient of t^n in $\varphi(n, t)$ is 1, since there is only one way to distribute the n distinct objects into n identical, or indistinguishable, boxes.

Example 12.9. From Table 12.1, or using Theorem 12.8, we see that the object distribution polynomial that counts the number of ways to distribute four distinct objects into identical boxes, with no box empty and ordered occupancy, is given by $\varphi(4, t) = 24t + 36t^2 + 12t^3 + t^4$. The coefficient of

t^3, for example, is 12. The 12 ordered distributions are shown below. (Remember, object order within a box matters, but box order does not matter for identical boxes.)

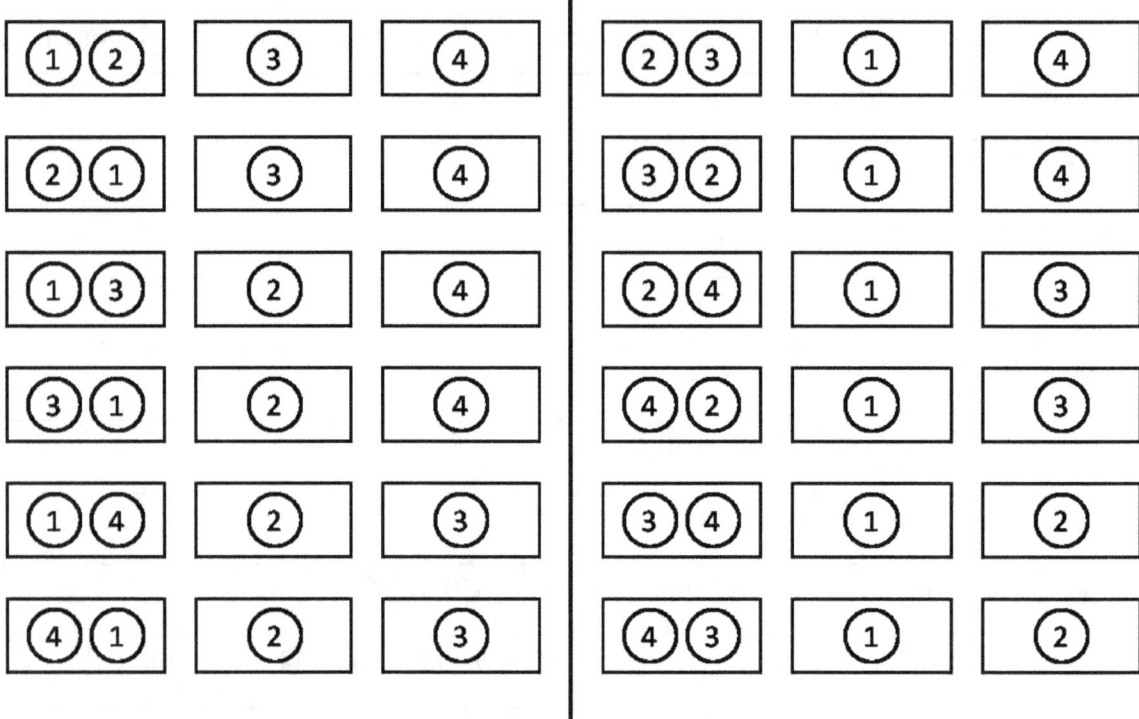

Figure 12.2. The 12 ordered distributions of Example 12.9.

Ordered Stirling Numbers of the Second Kind

In combinatorial mathematics one often encounters certain numbers that are called *Stirling numbers of the second kind*, denoted by $S(n, k)$. The numbers $S(n, k)$ count the number of ways to partition an *n*-set, say of *n* distinct objects, into precisely *k* non-empty disjoint subsets. By comparison, the coefficients of t^k counted by $\varphi(r, t)$ in Theorem 12.8, equation (12.13), are almost the Stirling numbers of the second kind, but with one important distinction. The numbers counted by Theorem 12.8 have *ordered* occupancy within each set. So, we can think of these numbers as *ordered Stirling numbers of the second kind*, and denote them as $S_O(r, k)$. These numbers are the coefficients of t^k in $\varphi(r, t)$:

$$\varphi(r, t) = \sum_{k=1}^{r} S_O(r, k) t^k .\qquad(12.14)$$

Corollary 12.10 (Ordered Stirling Numbers of the Second Kind). *A recurrence relation for ordered Stirling numbers of the second kind is given by*

$$S_O(r, k) = r \cdot S_O(r-1, k) + \frac{r}{k} \cdot S_O(r-1, k-1)\qquad(12.15)$$

where $S_O(n, n) = 1$ and $S_O(n, k) = 0$ if $n < k$.

Proof. Apply Theorem 12.8, equation (12.13) to (12.14) term-by-term. Then read off the coefficient of t^k which is $S_O(r, k)$. □

The recurrence relations in Theorem 12.8 and Corollary 12.10 are very useful. But suppose we did not have these recurrence relations. How could we calculate the numerical coefficients in $\varphi(r, t)$, or the numbers $S_O(r, k)$, directly? Consider, for example, the coefficient of t^2 in $\varphi(4, t)$, or $S_O(4, 2)$. We can calculate this number directly by looking at the possible types of object distributions into two boxes:

- Type A: { * } {***}, for example {1} {2, 3, 4}
- Type B: {**} {**}, for example {1, 2} {3, 4}

We used asterisks to represent the objects, but remember that the objects are *distinct*, like the numbers from 1 to 4. The partitions of 4 into 2 distinct parts are

- 4 = 1 + 3
- 4 = 2 + 2

So, to calculate $S_O(4, 2)$, which is 36 (Table 12.1), we can first permute the four distinct objects in 4! ways, then divide by the "overcount," because boxes containing the same number of objects can be permuted. We do this because the boxes themselves are identical; hence, box order is irrelevant. Then, for this example, we have $S_O(4, 2) = \frac{4!}{1!1!} + \frac{4!}{2!} = 24 + 12 = 36$. Here, the term $\frac{4!}{1!1!}$ counts the distributions corresponding to the partition 4 = 1 + 3 (i.e., the Type A distributions). Similarly, the term $\frac{4!}{2!}$ counts the distributions corresponding to the partition 4 = 2 + 2 (i.e., the Type B distributions).

We can formalize these calculations as follows. Let $\pi(r, k)$ denote the set of partitions of r into k parts. For a given partition, let k_j be the number of j-identical parts. For example, for the partition of $56 = 1 + 3 + 4 + 4 + 4 + 7 + 7 + 8 + 9 + 9$, we have $k_1 = 3$ (for 1, 3, and 8, which each occur once), $k_2 = 2$ (for 7 and 9, which each occur twice), and $k_3 = 1$ (for 4, which occurs three times). Using this notation, we can write

$$S_O(4, 2) = \sum_{\pi(4, 2)} \frac{4!}{1!^{k_1}\, 2!^{k_2}\, 3!^{k_3}\, 4!^{k_4}}. \tag{12.16}$$

Example 12.11. Let us calculate $S_O(5, 3)$. From Table 12.1, we know that $S_O(5, 3) = 120$. The partitions of 5 into 3 parts are $5 = 1 + 1 + 3$ and $5 = 1 + 2 + 2$. So, we have

$$S_O(5, 3) = \sum_{\pi(5, 3)} \frac{5!}{1!^{k_1}\, 2!^{k_2}\, 3!^{k_3}\, 4!^{k_4}\, 5!^{k_5}}$$

$$= \frac{5!}{1!\, 2!} + \frac{5!}{1!\, 2!}$$

$$= \frac{5!}{2} + \frac{5!}{2}$$

$$= 120.$$

Another notation, which is often used for integer partitions, is the following:

- $5 = 1 + 1 + 3 = [1^2 3^1] = [1^{e_1} 2^{e_2} 3^{e_3} ...]$. Here, $e_1 = 2, e_2 = 0, e_3 = 1$.
- $5 = 1 + 2 + 2 = [1^1 2^2] = [1^{e_1} 2^{e_2} 3^{e_3} ...]$. Here, $e_1 = 1, e_2 = 2, e_3 = 0$.

Then we have

$$S_O(5, 3) = \sum_{\pi(5, 3)} \binom{5}{e_1, e_2, ...}$$

$$= \binom{5}{2, 0, 1} + \binom{5}{1, 2, 0}$$

$$= \binom{5}{2, 1} + \binom{5}{1, 2}$$

$$= \frac{5!}{2!1!} + \frac{5}{1!2!}$$

$$= 60 + 60$$

$$= 120.$$

Corollary 12.10 gives us a recurrence relation for $S_O(r, k)$, namely

$S_O(r, k) = r \cdot S_O(r-1, k) + \dfrac{r}{k} \cdot S_O(r-1, k-1)$. How can we solve this recurrence relation to find an explicit formula for $S_O(r, k)$? Consider, for example, $S_O(7, 3)$. Using the recurrence, we obtain

$S_O(7, 3) = 7 \cdot S_O(6, 3) + \dfrac{7}{3} \cdot S_O(6, 2)$. Next, we can use the recurrence again, by back-substitution, for $S_O(6, 3)$ and $S_O(6, 2)$. Then we can use back-substitution again, and so on. Continuing in this way, we will ultimately obtain $S_O(7, 3) = 15 \cdot \dfrac{7!}{3!}$. At this point, we use mathematical intuition based on experience. The number 15 is not a coincidence. It must mean something. Since this is a combinatorial problem, the number 15 should be something combinatorial. We should always look for our combinatorial numbers in Pascal's triangle. A natural guess is that $15 = \binom{6}{2}$. So,

$S_O(7, 3) = \binom{6}{2} \dfrac{7!}{3!}$. Therefore, we conjecture that $S_O(r, k) = \dfrac{r!}{k!} \binom{r-1}{k-1}$.

Theorem 12.12. $S_O(r, k) = \dfrac{r!}{k!} \binom{r-1}{k-1}$.

Proof. We will show that this formula for $S_O(r, k)$ satisfies both the recurrence relation

$S_O(r, k) = r \cdot S_O(r-1, k) + \dfrac{r}{k} \cdot S_O(r-1, k-1)$ and the initial conditions $S_O(r, 1) = r!$ and $S_O(r, r) = 1$.

Substituting $S_O(r, k) = \dfrac{r!}{k!} \binom{r-1}{k-1}$ into the recurrence relation, we obtain

$$\binom{r-1}{k-1} \dfrac{r!}{k!} = r \binom{r-2}{k-1} \dfrac{(r-1)!}{k!} + \dfrac{r}{k} \binom{r-2}{k-2} \dfrac{(r-1)!}{(k-1)!}$$

$$= \binom{r-2}{k-1} \dfrac{r!}{k!} + \binom{r-2}{k-2} \dfrac{r!}{k!}$$

This implies that $\binom{r-1}{k-1} = \binom{r-2}{k-1} + \binom{r-2}{k-2}$, which is true by Pascal's identity. Next, let's check the initial conditions. First, note that $S_O(r, 1) = \binom{r-1}{0}\frac{r!}{1!} = r!$. Finally, note that $S_O(r, r) = \binom{r-1}{r-1}\frac{r!}{r!} = 1$. Thus, our formula satisfies both the recurrence relation and the initial conditions. This proves the theorem. □

Corollary 12.13. $\sum_{\pi(r,k)} \binom{k}{e_1, e_2, ..., e_r} = \binom{r-1}{k-1}$, where we are summing over the partitions of r into k parts, and the e_j's are the exponents in the partition notation for r for each partition (i.e., $[1^{e_1} 2^{e_2} ... r^{e_r}]$).

Proof. Since $S_O(r, k) = \sum_{\pi(r,k)} \binom{r}{e_1, e_2, ..., e_r}$ and $S_O(r, k) = \frac{r!}{k!}\binom{r-1}{k-1}$, equating these two expressions and simplifying gives the desired result. □

Corollary 12.13 relates a sum of multinomial coefficients over integer partitions to a binomial coefficient: The sum over the partitions of r into k parts of the multinomial coefficients of k in the part repetitions is equal to the number of combinations of $r - 1$ things chosen $k - 1$ at a time:

$$\sum_{\text{partitions of } r \text{ into } k \text{ parts}} \binom{k}{\text{part repetitions}} = \binom{r-1}{k-1}.$$

Example 12.14. The partitions of 7 into 4 parts are
- $7 = 4 + 1 + 1 + 1 = [1^3 4^1]$
- $7 = 3 + 2 + 1 + 1 = [1^2 2^1 3^1]$
- $7 = 2 + 2 + 2 + 1 = [1^1 2^3]$

Then we have the following:

$$\sum_{\pi(7,4)}\binom{4}{e_1, e_2, ..., e_7} = \binom{4}{3, 1} + \binom{4}{2, 1, 1} + \binom{4}{1, 3}$$
$$= \frac{4!}{3!1!} + \frac{4!}{2!1!1!} + \frac{4!}{1!3!}$$
$$= 4 + 12 + 4$$
$$= 20.$$

Also, $\binom{r-1}{k-1} = \binom{7-1}{4-1} = \binom{6}{3} = 20$.

Exercises

1. Use Theorem 12.5 to calculate the object distribution polynomial $f_n(t)$ for a set of objects consisting of two red balls and two blue balls.

2. Use Theorem 12.8 to recursively calculate $\varphi(1, t)$, $\varphi(2, t)$, $\varphi(3, t)$, and $\varphi(4, t)$. Verify your results against Table 12.1.

3. Use Corollary 12.10 to recursively calculate $S_O(6, 4)$. Compare your results with the coefficient of t^4 in $\varphi(6, t)$, Tabl 12.1.

4. Verify Corollary 12.13 for the positive integer partitions of 7 into 3 parts. What are the partitions of 7 into 3 parts? List them.

222

Chapter 13. Algebraic Identities for Object Distribution Polynomials

In the previous chapters we focused our attention on developing methods to calculate object distribution polynomials. Since explicit formulas only exist for special cases, we instead developed several recurrence relations to calculate these polynomials. In this chapter we will focus on the algebraic properties of object distribution polynomials. There are deep, fundamental relationships among the various object distribution problems, and these relationships are most elegantly expressed in the form of algebraic identities among the object distribution polynomials. Algebraic identities appear frequently in mathematics. In trigonometry, for example, we have the well-known identity $\sin^2 x + \cos^2 x = 1$, which holds true for any real number x. Algebraic identities, in all branches of mathematics, are generally beautiful, important, and useful. We will state and prove several algebraic identities that were discovered by the author of this book.

In order to derive our algebraic identities, we will need several generating functions. Here, we will simply state the generating functions. Deriviations of these generating functions were presented in previous chapters.

The enumeration logarithmic generating function for counting the number of ways to distribute objects of specification n into identical boxes with no box empty is given by

$$G(x, t) = \sum_{n=1}^{\infty} g_n(t) x^{\log n} = \prod_{d>1} \frac{1}{(1 - tx^{\log d})}. \qquad (13.1)$$

The polynomial $g_n(t)$ counts the number of distributions of objects into identical boxes with no box empty. We are also interested in object distribution polynomials that count the number of distributions into identical boxes having distinct occupancy, $\delta_n(t)$, and the distributions having even color, or even type, occupancy, $\varepsilon_n(t)$.

Let $\delta_n(t)$ be the object distribution polynomial for distributing colored balls of specification n into identical boxes with distinct occupancies and no box empty. The logarithmic generating function for $\delta_n(t)$ is given by

$$\Delta(x,t) = \prod_{d>1}(1+tx^{\log d}) = \sum_{n=1}^{\infty} \delta_n(t) x^{\log n}. \tag{13.2}$$

Let $\varepsilon_n(t)$ be the object distribution polynomial for distributing colored balls of specification n into identical boxes, with no boxes empty, such that each color that occurs within a given box occurs with even replication. The logarithmic generating function for $\varepsilon_n(t)$ is given by

$$E(x,t) = \prod_{d>1} \frac{1}{(1-tx^{2\log d})} = \sum_{n=1}^{\infty} \varepsilon_n(t) x^{\log n}. \tag{13.3}$$

We now state and prove several algebraic identities for these object distribution polynomials.

Theorem 13.1. *Let $g_n(t)$ be the object distribution polynomial that counts the number of distributions of objects, of specification n, into identical boxes with no box empty. Then*

$$g_{n^2}(t) * g_{n^2}(-t) = g_n(t^2). \tag{13.4}$$

Proof. The theorem is easily proved using the logarithmic generating function, equation (13.1), as follows.

$$\sum_{n=1}^{\infty} g_n(t) x^{\log n} = \prod_{d>1} \frac{1}{(1-tx^{\log d})}$$

$$\sum_{n=1}^{\infty} g_n(t^2)(x^2)^{\log n} = \prod_{d>1} \frac{1}{(1-t^2(x^2)^{\log d})}$$

$$\sum_{n=1}^{\infty} g_n(t^2) x^{\log n^2} = \prod_{d>1} \frac{1}{(1-t^2(x^{\log d})^2)}$$

$$= \prod_{d>1} \frac{1}{(1-tx^{\log d})(1+tx^{\log d})}$$

$$= \prod_{d>1} \frac{1}{(1-tx^{\log d})} \prod_{d>1} \frac{1}{(1+tx^{\log d})}$$

$$= \sum_{n=1}^{\infty} g_n(t) x^{\log n} \sum_{n=1}^{\infty} g_n(-t) x^{\log n}$$

$$= \sum_{n=1}^{\infty} \left(g_n(t) * g_n(-t) \right) x^{\log n}$$

Therefore, we have

$$\sum_{n=1}^{\infty} g_n(t^2) x^{\log n^2} = \sum_{n=1}^{\infty} \left(g_n(t) * g_n(-t) \right) x^{\log n} \qquad (13.5)$$

From the left-hand side of (13.5), we observe that the coefficient of $x^{\log n^2}$ is $g_n(t^2)$. And from the right-hand side of (13.5), the coefficient of $x^{\log n^2}$ is $g_{n^2}(t) * g_{n^2}(-t)$. Therefore, by equating coefficients of $x^{\log n^2}$, we have $g_{n^2}(t) * g_{n^2}(-t) = g_n(t^2)$. □

Comment 1. We sometimes refer to identity (13.4) as a "square-root" identity because it can be written in the following form (provided, of course, that n is perfect square):

$$g_n(\sqrt{t}) * g_n(-\sqrt{t}) = g_{\sqrt{n}}(t). \qquad (13.6)$$

Comment 2. The "product" $*$ is the standard convolution product from number theory. So, $g_{n^2}(t) * g_{n^2}(-t) = g_n(t^2)$ means $\sum_{d|n^2} g_{\frac{n^2}{d}}(t) g_d(-t) = g_n(t^2)$.

Comment 3. The convolution product is commutative: $f(n) * h(n) = h(n) * f(n)$. This is proved in any standard text on number theory.

Theorem 13.2. Let $\delta_n(t)$ be the object distribution polynomial that counts the number of distributions of objects, of specification n, into identical boxes, with no box empty, such that the box occupancies are distinct. Then

$$\delta_{n^2}(t) * \delta_{n^2}(-t) = \delta_n(-t^2). \tag{13.7}$$

Proof. We use the distinct occupancy generating function, (13.2), and follow the same method as in the proof of Theorem 13.1.

$$\sum_{n=1}^{\infty} \delta_n(t) x^{\log n} = \prod_{d>1} (1+tx^{\log d})$$

$$\sum_{n=1}^{\infty} \delta_n(-t^2)(x^2)^{\log n} = \prod_{d>1} (1-t^2(x^2)^{\log d})$$

$$\sum_{n=1}^{\infty} \delta_n(-t^2) x^{\log n^2} = \prod_{d>1} (1-t^2(x^{\log d})^2)$$

$$= \prod_{d>1} (1-tx^{\log d})(1+tx^{\log d})$$

$$= \prod_{d>1} (1-tx^{\log d}) \prod_{d>1} (1+tx^{\log d})$$

$$= \sum_{n=1}^{\infty} \delta_n(-t) x^{\log n} \sum_{n=1}^{\infty} \delta_n(t) x^{\log n}$$

$$= \sum_{n=1}^{\infty} \left(\delta_n(-t) * \delta_n(t) \right) x^{\log n}$$

Therefore, we have

$$\sum_{n=1}^{\infty} \delta_n(-t^2) x^{\log n^2} = \sum_{n=1}^{\infty} \left(\delta_n(-t) * \delta_n(t) \right) x^{\log n} \tag{13.8}$$

From the left-hand side of (13.8), we observe that the coefficient of $x^{\log n^2}$ is $\delta_n(-t^2)$. And from the right-hand side of (13.8), the coefficient of $x^{\log n^2}$ is $\delta_{n^2}(t) * \delta_{n^2}(-t)$. Therefore, by equating coefficients of $x^{\log n^2}$, we have $\delta_{n^2}(t) * \delta_{n^2}(-t) = \delta_n(-t^2)$. □

Theorem 13.3. *Let $\varepsilon_n(t)$ be the object distribution polynomial that counts the number of distributions of objects, of specification n, into identical boxes, with no box empty, such that each box has even color, or even type, occupancy. Then*

$$\varepsilon_{n^2}(t) * \varepsilon_{n^2}(-t) = \varepsilon_n(t^2). \tag{13.9}$$

Proof. We use the logarithmic generating function for even color occupancy, (13.3).

$$\sum_{n=1}^{\infty} \varepsilon_n(t) x^{\log n} = \prod_{d>1} \frac{1}{(1-tx^{2\log d})}$$

$$\sum_{n=1}^{\infty} \varepsilon_n(t^2)(x^2)^{\log n} = \prod_{d>1} \frac{1}{(1-t^2(x^2)^{2\log d})}$$

$$\sum_{n=1}^{\infty} \varepsilon_n(t^2) x^{\log n^2} = \prod_{d>1} \frac{1}{(1-t^2(x^2)^{2\log d})}$$

$$= \prod_{d>1} \frac{1}{(1-tx^{2\log d})(1+tx^{2\log d})}$$

$$= \prod_{d>1} \frac{1}{(1-tx^{2\log d})} \prod_{d>1} \frac{1}{(1+tx^{2\log d})}$$

$$= \sum_{n=1}^{\infty} \varepsilon_n(t) x^{\log n} \sum_{n=1}^{\infty} \varepsilon_n(-t) x^{\log n}$$

$$= \sum_{n=1}^{\infty} \left(\varepsilon_n(t) * \varepsilon_n(-t) \right) x^{\log n}$$

Therefore, we have

$$\sum_{n=1}^{\infty} \varepsilon_n(t^2) x^{\log n^2} = \sum_{n=1}^{\infty} \left(\varepsilon_n(t) * \varepsilon_n(-t) \right) x^{\log n} \tag{13.10}$$

From the left-hand side of (13.10), we observe that the coefficient of $x^{\log n^2}$ is $\varepsilon_n(t^2)$. And from the right-hand side of (13.10), the coefficient of $x^{\log n^2}$ is $\varepsilon_{n^2}(t) * \varepsilon_{n^2}(-t)$. Therefore, by equating the coefficients of $x^{\log n^2}$, we have $\varepsilon_{n^2}(t) * \varepsilon_{n^2}(-t) = \varepsilon_n(t^2)$. □

Theorem 13.4.

$$g_n(t) * g_n(-t) = \varepsilon_n(t^2) \tag{13.11}$$

Proof.

$$\sum_{n=1}^{\infty} \varepsilon_n(t^2) x^{\log n} = \prod_{d>1} \frac{1}{(1-t^2 x^{2\log d})}$$

$$= \prod_{d>1} \frac{1}{(1-tx^{\log d})(1+tx^{\log d})}$$

$$= \prod_{d>1} \frac{1}{(1-tx^{\log d})} \prod_{d>1} \frac{1}{(1+tx^{\log d})}$$

$$= \sum_{n=1}^{\infty} g_n(t) x^{\log n} \sum_{n=1}^{\infty} g_n(-t) x^{\log n}$$

$$= \sum_{n=1}^{\infty} \left(g_n(t) * g_n(-t) \right) x^{\log n}$$

Equating the coefficients of $x^{\log n}$ on both sides of the equation gives $g_n(t) * g_n(-t) = \varepsilon_n(t^2)$. $\square$

Theorem 13.5. *Every object distribution polynomial $g_n(t)$ is also an even color occupancy object distribution polynomial:*

$$g_n(t) = \varepsilon_{n^2}(t). \tag{13.12}$$

Proof. By Theorem 13.1, $g_{n^2}(t) * g_{n^2}(-t) = g_n(t^2)$. Then, by Theorem 13.4, with n replaced by n^2, we have $g_{n^2}(t) * g_{n^2}(-t) = \varepsilon_{n^2}(t^2)$. Hence, $g_n(t^2) = \varepsilon_{n^2}(t^2)$, or simply $g_n(u) = \varepsilon_{n^2}(u)$ upon change of variables, say $u = t^2$. $\square$

Although Theorem 13.5 was proved algebraically, it is fairly obvious after a moment of thought. Given any object distribution counted by $g_n(t)$, if we simply double the set of objects in each box we will obtain a distribution for even color occupancy, having specification number n^2, which is counted by $\varepsilon_{n^2}(t)$. Conversely, given any distribution having even color occupancy counted by $\varepsilon_{n^2}(t)$, if we halve the objects of each color, within each box, we will obtain an object distribution that is counted by $g_n(t)$.

Theorem 13.6. *Let $g_n(t)$ be an object distribution polynomial. Then we have*

$$g_n(t) * g_n(-t) = \begin{cases} g_m(t^2), & \text{if } n = m^2, \\ 0, & \text{if } n \neq m^2 \end{cases} \quad (13.13)$$

for some positive integer m.

Proof. First, if $n = m^2$, for some positive integer m, then, by Theorem 13.1, $g_n(t) * g_n(-t) = g_{m^2}(t) * g_{m^2}(-t) = g_m(t^2)$. Next, if $n \neq m^2$, for some positive integer m, then by Theorem 13.4, $g_n(t) * g_n(-t) = \varepsilon_n(t^2) = 0$, because $\varepsilon_n(t^2)$ is nonzero if and only if n is a perfect square. Why? An object distribution can have even color occupancy if and only if there are an even number of objects of each color. This means that the specification number n must have the form

$$n = p_1^{2a_1} p_2^{2a_2} \ldots p_r^{2a_r} = (p_1^{a_1} p_2^{a_2} \ldots p_r^{a_r})^2. \quad \square$$

Example 13.7. Let $n = 12 = 2^2 \cdot 3$. Here, 12 is not a perfect square, so Theorem 13.6 tells us that $g_{12}(t) * g_{12}(-t) = 0$. Let's check this:

$$\begin{aligned} g_{12}(t) * g_{12}(-t) &= \sum_{d|12} g_d(t) g_{\frac{12}{d}}(-t) \\ &= g_1(t)g_{12}(-t) + g_2(t)g_6(-t) + g_3(t)g_4(-t) + g_4(t)g_3(-t) \\ &\quad + g_6(t)g_2(-t) + g_{12}(t)g_1(-t) \\ &= 1 \cdot (-t + 2t^2 - t^3) + t(-t + t^2) + t(-t + t^2) + (t + t^2)(-t) \\ &\quad + (t + t^2)(-t) + (t + 2t^2 + t^3) \cdot 1 \\ &= 0. \end{aligned}$$

Let us remind ourselves of a basic definition:

Definition 13.8. An object distribution polynomial, $g_n(t)$, counts the number of ways to distribute objects of specification n into identical boxes with no box empty. The coefficient of t^k in $g_n(t)$ is the number of such distributions into precisely k identical boxes with no box empty.

Example 13.9. Suppose we have two red balls, one white ball, and one blue ball. We can represent this set of objects by the multiset $S = \{r, r, w, b\}$. To encode this set of objects in a *specification number n*, we assign a distinct prime number to each color. Let's assign the prime 2 to "red," 3 to "white," and 5 to "blue." Then, the specification number is $n = 2^2 \cdot 3 \cdot 5 = 60$. The object distribution polynomial is given by $g_{60}(t) = t + 5t^2 + 4t^3 + t^4$. If we had used a different set of prime numbers, such as 3, 7, and 11 for red, white, and blue, respectively, our specification number would be $n = 3^2 \cdot 7 \cdot 11 = 693$. In that case, we still would obtain the *same* object distribution polynomial, namely $g_{693}(t) = t + 5t^2 + 4t^3 + t^4$. In other words, $g_{60}(t) = g_{693}(t)$. This is because the specification numbers 60 and 693 have the same *form* when they are prime factorized as $p_1^2 \cdot p_2 \cdot p_3$, for distinct primes p_1, p_2, and p_3. The coefficient of, say, t^3 is 4, and this means that there are four different ways to distribute two red balls, one white ball, and one blue ball into precisely three identical boxes with no box empty.

When we have boxes of several types, or "colors," then the object distribution polynomials count the number of distributions of objects, of specification n, into all the various combinations of, say, k boxes selected from a set of r colors of boxes. Naturally, the object distribution polynomials for boxes, or cells, of mixed type (colors) are more complicated than the polynomials for identical, or indistinguishable, boxes. Nevertheless, the mathematical theory exists for dealing with these cases.

Definition 13.10. An object distribution polynomial of the form $g_n(t, \alpha_1, \alpha_2, \ldots, \alpha_r)$ counts the number of ways to distribute objects of specification n into boxes chosen from a set of r colors, with color repetitions allowed, and with no box empty. The coefficient of t^k is the number of such distributions into precisely k boxes with no box empty.

Example 13.11. Suppose we have a multiset of objects, such as colored balls, given by $S = \{r, r, b\}$ for red, red, and blue. That is, we have two identical red balls and one blue ball. For a specification number, let's use the prime 2 for a red ball and 3 for a blue ball. We then have $n = 2 \cdot 2 \cdot 3 = 12$. If we have a supply of boxes that come in two colors, say black and white, then the object distribution polynomial (see Chapter 9) turns out to be the following expression:

$$g_{12}(t, \alpha, \beta) = (\alpha + \beta)t + (2\alpha^2 + 4\alpha\beta + 2\beta^2)t^2 + (\alpha^3 + 2\alpha^2\beta + 2\alpha\beta^2 + \beta^3)t^3.$$

Here, we let α = a black box and β = a white box. For object distributions into precisely two boxes, we look at the coefficient of t^2 in $g_{12}(t, \alpha, \beta)$, which is $2\alpha^2 + 4\alpha\beta + 2\beta^2$. The term $4\alpha\beta$, for example, tells us that there are four different distributions of the objects into a box of color α (black) and a box of color β (white) as follows:

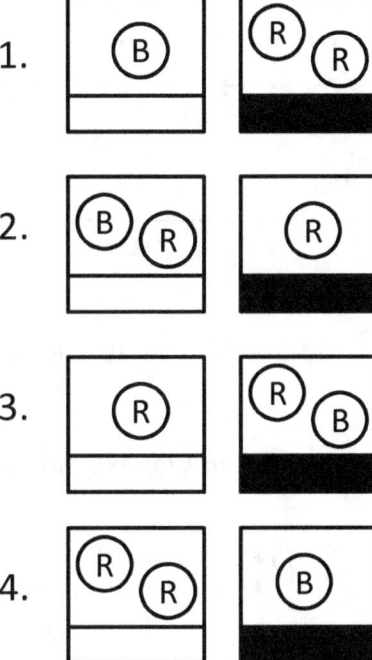

For the class of distribution problems that we are considering, the *order* of the boxes does not matter. Also, all of the boxes of a given color are identical. Similarly, all of the objects of a given color are identical.

Table 13.1 presents several examples of general object distribution polynomials for boxes of two colors (color α and color β). We will use the polynomials in Table 13.1 in later examples. These polynomials are calculated using the theory presented in Chapter 9 of this book.

Table 13.1. Object Distribution Polynomials for Boxes of Two Colors

Object Distribution Polynomial	Notes
$g_1(t, \alpha, \beta) = 1$	By definition
$g_p(t, \alpha, \beta) = (\alpha + \beta)t$	Prime p
$g_{p^2}(t, \alpha, \beta) = (\alpha + \beta)t + (\alpha^2 + \alpha\beta + \beta^2)t^2$	Prime p
$g_{pq}(t, \alpha, \beta) = (\alpha + \beta)t + (\alpha^2 + 2\alpha\beta + \beta^2)t^2$	Distinct primes p and q
$g_{p^2q}(t, \alpha, \beta) = (\alpha + \beta)t + (2\alpha^2 + 4\alpha\beta + 2\beta^2)t^2 + (\alpha^3 + 2\alpha^2\beta + 2\alpha\beta^2 + \beta^3)t^3$	Distinct primes p and q

Now we want to generalize Theorem 13.6, which we restate for reference:

Theorem 13.12. *Let $g_n(t)$ be an object distribution polynomial. Then we have*

$$g_n(t) * g_n(-t) = \begin{cases} g_m(t^2), & \text{if } n = m^2, \\ 0, & \text{if } n \neq m^2 \end{cases}$$

for some positive integer m.

In Theorem 13.12, the convolution product ($*$) is the standard number-theoretic product defined by $f(n) * g(n) = \sum_{d|n} f(d) g(n/d)$ for positive integer n. We now want to generalize Theorem 13.12. We will begin with a special case and an illustrative example. Then we will state and prove the generalized version of Theorem 13.12.

Definition 13.13. If we are only interested in counting the total number of object distributions into boxes selected from a supply of r colors, without actually specifying the distributions by type, then we can simplify the object distribution polynomial $g_n(t, \alpha_1, \alpha_2, \ldots, \alpha_r)$ by setting $\alpha_1 = \alpha_2 = \ldots = \alpha_r = 1$. We then define $g_n(t, r) = g_n(t, 1, 1, \ldots, 1)$.

Theorem 13.14. *If n is not a perfect square, then* $g_n(t, r) * g_n(-t, r) = 0$.

Proof. Based on the theory developed in Chapter 9, we have
$g_n(t, \alpha_1, \alpha_2, \ldots, \alpha_r) = g_n(\alpha_1 t) * g_n(\alpha_2 t) * \ldots * g_n(\alpha_r t)$. Then, by Definition 13.13 and Theorem 13.12, we have

$$g_n(t, r) * g_n(-t, r) = \big(g_n(t) * \ldots * g_n(t)\big) * \big(g_n(-t) * \ldots * g_n(-t)\big)$$
$$= \big(g_n(t) * \ldots * g_n(t)\big) * \big(g_n(t) * g_n(-t)\big) * \big(g_n(-t) * \ldots * g_n(-t)\big)$$
$$= \big(g_n(t) * \ldots * g_n(t)\big) * 0 * \big(g_n(-t) * \ldots * g_n(-t)\big)$$
$$= 0. \quad \square$$

Example 13.15. Let $n = 12 = 2^2 \cdot 3$. For boxes of two colors, say α and β, Table 13.1 gives

$g_{12}(t, \alpha, \beta) = (\alpha + \beta)t + (2\alpha^2 + 4\alpha\beta + 2\beta^2)t^2 + (\alpha^3 + 2\alpha^2\beta + 2\alpha\beta^2 + \beta^3)t^3$. Then

$g_{12}(t, 2) = g_{12}(t, 1, 1) = 2t + 8t^2 + 6t^3$. Since 12 is not a perfect square, Theorem 13.14 tells us that

$g_{12}(t, 2) * g_{12}(-t, 2) = 0$. Let's verify this using the polynomials from Table 13.1 with $\alpha = \beta = 1$. We have

$$g_{12}(t, 2) * g_{12}(-t, 2) = \sum_{d \mid 12} g_d(t, 2) g_{\frac{12}{d}}(-t, 2)$$
$$= g_1(t, 2)g_{12}(-t, 2) + g_2(t, 2)g_6(-t, 2) + g_3(t, 2)g_4(-t, 2)$$
$$+ g_4(t, 2)g_3(-t, 2) + g_6(t, 2)g_2(-t, 2) + g_{12}(t, 2)g_1(-t, 2)$$
$$= (1)(-2t + 8t^2 - 6t^3) + (2t)(-2t + 4t^2) + (2t)(-2t + 3t^2)$$
$$+ (2t + 3t^2)(-2t) + (2t + 4t^2)(-2t) + (2t + 8t^2 + 6t^3)(1)$$
$$= 0. \quad \square$$

We are now prepared to state and prove the general convolution identity for object distribution polynomials. Theorem 13.16 is a generalization of Theorem 13.12.

Theorem 13.16 (General Convolution Product Identity). *Let $g_n(t, \alpha_1, \alpha_2, \ldots, \alpha_r)$ be the object distribution polynomial that counts the number of ways to distribute objects of specification n into any combination of colored boxes, selected from an unlimited supply of r colors, with no box empty. We have the following convolution product identity:*

$$g_n(t, \alpha_1, \alpha_2, \ldots, \alpha_r) * g_n(-t, \alpha_1, \alpha_2, \ldots, \alpha_r) = \begin{cases} g_m(t^2, \alpha_1^2, \alpha_2^2, \ldots, \alpha_r^2) & \text{if } n = m^2, \\ 0, & \text{if } n \neq m^2 \end{cases}$$

for some integer m.

Proof. The proof uses Theorem 13.12 together with the fact that the convolution product is commutative and associative. We have $g_n(t, \alpha_1, \alpha_2, \ldots, \alpha_r) = g_n(\alpha_1 t) * g_n(\alpha_2 t) * \ldots * g_n(\alpha_r t)$. We consider two cases:

Case 1. If $n \neq m^2$, for positive integer m, then

$$g_n(t, \alpha_1, \alpha_2, \ldots, \alpha_r) * g_n(-t, \alpha_1, \alpha_2, \ldots, \alpha_r)$$
$$= \left(g_n(\alpha_1 t) * \ldots * g_n(\alpha_r t)\right) * \left(g_n(-\alpha_1 t) * \ldots * g_n(-\alpha_r t)\right)$$
$$= (g_n(\alpha_r t) * \ldots * g_n(\alpha_2 t)) * (g_n(\alpha_1 t) * g_n(-\alpha_1 t)) * (g_n(-\alpha_2 t) * \ldots * g_n(-\alpha_r t))$$
$$= (g_n(\alpha_r t) * \ldots * g_n(\alpha_2 t)) * 0 * (g_n(-\alpha_2 t) * \ldots * g_n(-\alpha_r t))$$
$$= 0.$$

Here, we used Theorem 13.12 for $g_n(\alpha_1 t) * g_n(-\alpha_1 t) = 0$ for n not a perfect square.

Case 2. If $n = m^2$ for positive integer m, then

$$g_n(t, \alpha_1, \alpha_2, \ldots, \alpha_r) * g_n(-t, \alpha_1, \alpha_2, \ldots, \alpha_r)$$
$$= \left(g_n(\alpha_1 t) * \ldots * g_n(\alpha_r t)\right) * \left(g_n(-\alpha_1 t) * \ldots * g_n(-\alpha_r t)\right)$$
$$= (g_n(\alpha_r t) * \ldots * g_n(\alpha_2 t)) * (g_n(\alpha_1 t) * g_n(-\alpha_1 t)) * (g_n(-\alpha_2 t) * \ldots * g_n(-\alpha_r t))$$
$$= (g_n(\alpha_r t) * \ldots * g_n(\alpha_2 t)) * (g_m(\alpha_1^2 t^2)) * (g_n(-\alpha_2 t) * \ldots * g_n(-\alpha_r t))$$
$$= g_m(\alpha_1^2 t^2) * (g_n(\alpha_r t) * \ldots * g_n(\alpha_3 t)) * (g_n(\alpha_2 t) * g_n(-\alpha_2 t)) * (g_n(-\alpha_3 t) * \ldots * g_n(-\alpha_r t))$$
$$= g_m(\alpha_1^2 t^2) * (g_n(\alpha_r t) * \ldots * g_n(\alpha_3 t)) * (g_m(\alpha_2^2 t^2)) * (g_n(-\alpha_3 t) * \ldots * g_n(-\alpha_r t))$$
$$= \ldots$$
$$= g_m(\alpha_1^2 t^2) * g_m(\alpha_2^2 t^2) * \ldots * g_m(\alpha_r^2 t^2)$$
$$= g_m(t^2, \alpha_1^2, \alpha_2^2, \ldots, \alpha_r^2). \quad \square$$

Now let's look at an example of Theorem 13.16 using the object distribution polynomials in Table 13.1.

Example 13.17. Let $n = 4$, which is a perfect square. Using the object distribution polynomials provided in Table 13.1, we can calculate $g_4(t, \alpha, \beta) * g_4(-t, \alpha, \beta)$ and see if we get $g_2(t^2, \alpha^2, \beta^2)$ as required by Theorem 13.16. We have

$$g_4(t, \alpha, \beta) * g_4(-t, \alpha, \beta) = \sum_{d|4} g_d(t, \alpha, \beta) g_{\frac{4}{d}}(-t, \alpha, \beta)$$

$$= g_1(t, \alpha, \beta)g_4(-t, \alpha, \beta) + g_2(t, \alpha, \beta)g_2(-t, \alpha, \beta) + g_4(t, \alpha, \beta)g_1(-t, \alpha, \beta)$$

$$= 1 \cdot \left(-(\alpha+\beta)t + (\alpha^2 + \alpha\beta + \beta^2)t^2\right) + \left((\alpha+\beta)t\right)\left(-(\alpha+\beta)t\right)$$

$$+ \left((\alpha+\beta)t + (\alpha^2 + \alpha\beta + \beta^2)t^2\right) \cdot 1$$

$$= (\alpha^2 + \beta^2)t^2$$

$$= g_2(t^2, \alpha^2, \beta^2). \quad \square$$

Exercises

1. Verify Theorem 13.1 for $n = 6$.
2. Verify Theorem 13.2 for $n = 6$.
3. Use the polynomials in Appendix A and Theorem 13.4 to show that $\varepsilon_2(t^2) = 0$. Equivalently, $\varepsilon_2(u) = 0$ for $u = t^2$. Explain what this means in terms of distributing objects into boxes.
4. Verify Theorem 13.6 for $n = 4$ and $n = 12$.

Chapter 14. A Generalization of Polya's Enumeration Theorem

The main goal of this chapter is to show how the number theoretic method can be used to extend and generalize Polya's Enumeration Theorem. Polya's Enumeration Theorem, also known as the Redfield-Polya theorem, is an important theorem in enumerative combinatorics. Polya's theorem can be used to count the number of nonequivalent configurations of some mathematical object, structure, or set taking symmetries into account. A simple example is the problem of counting the number of necklaces that we can create using, say, four black beads and three white beads. We could certainly count the number of circular permutations of four black beads and three white beads, but then we must take the various symmetries into account to avoid double counting. This is because the necklace can be rotated or flipped over. Two necklace configurations that, on first glance, appear to be distinct may actually be identical, because it may be possible to rotate or flip one configuration into the other. Technically, we need to take into account the group of dihedral symmetries of the necklace. This is just one example where Polya's counting theorem allows us to mechanize, or automate, the counting problem.

Polya's theorem is a unification, or synthesis, of mathematical group theory (a branch of abstract algebra) and generating functions. Group theory is used to account for all the symmetries of an object, which helps to identify equivalent configurations. Generating functions are used to systematically calculate the number of nonequivalent configurations. Combining these two great ideas produces a powerful combinatorial enumeration tool—*Polya's Enumeration Theorem*.

A complete and thorough treatment of Polya's Enumeration Theorem requires considerable preparation in preliminary topics such as abstract algebra (e.g., equivalence relations and group theory) and generating functions. Since our goal is to show how we can generalize Polya's theorem using the number theoretic method, we will not provide a complete and rigorous development of Polya's theory. (The reader should consult standard advanced textbooks on enumerative combinatorics, such as [2] and [5].) Instead, we will present an overview and summary of the theory, assuming that the reader is already familiar with group theory and generating functions, and a concrete example to illustrate how Polya's theorem is used to solve an enumeration problem. Then we will state and prove the author's generalization of Polya's theorem using the number theoretic method.

Polya's Enumeration Theorem

Theorem 14.1 (Polya's Enumeration Theorem). *Let X be a set of elements and G a group of permutations of X that acts to induce an equivalence relation on the colorings of X. The inventory of nonequivalent colorings of X using m colors $c_1, c_2, \ldots, c_m$ is given by the generating function*

$$P_G(x_1, \ldots, x_k) = P_G\left(\sum_{j=1}^{m} c_j, \sum_{j=1}^{m} c_j^2, \ldots, \sum_{j=1}^{m} c_j^k\right),$$ *where the cycle index for the group G of symmetries acting on the set X is given by* $P_G(x_1, \ldots, x_k) = \dfrac{1}{|G|} \sum_{\pi \in G} x_1^{j_1(\pi)} \ldots x_k^{j_k(\pi)}.$

We will not prove Polya's theorem because that would require an excursion into group theory, which is beyond the scope of this book. For a proof of the theorem, see [2]. Although Polya's theorem appears rather complicated on first examination, it is actually quite easy to use in practice. A simple example will illustrate how to use Polya's theorem to solve an enumeration problem.

Example 14.2. Determine the number of nonequivalent 3-bead necklaces that can be created using black (*b*) and white (*w*) beads taking rotational symmetries (but not flips) into account. While it would be simple and straightforward to calculate the number of permutations of various combinations of black and white beads, the complication is that we want to account for the various rotational symmetries of the necklace. (In this example, we will not worry about flips, but Polya's theorem can easily handle that, too.) Two necklaces that appear, on first glance, to be distinct may be identical if we can rotate one necklace into the other. We will solve this problem using Polya's counting theorem. We model the 3-bead necklace as an equilateral triangle as shown in Figure 14.1.

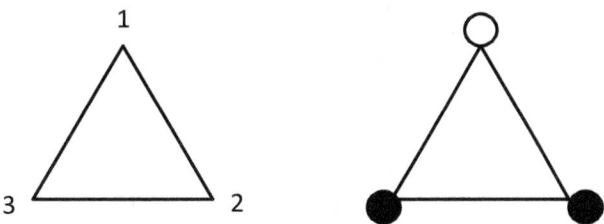

Figure 14.1. Equilateral triangle and a 3-bead necklace.

Before we can use Polya's counting theorem, we must determine the cycle index P_G for the symmetry group G that permutes the vertices of the equilateral triangle. We are only concerned with rigid mappings of the equilateral triangle into itself by rotations. So we can have rotations of 0, 120, or 240 degrees. A rotation of zero degrees is the identity element of the group. Thus, the symmetry group is $G = \{\pi_0, \pi_{120}, \pi_{240}\}$. Since G has three distinct elements, its order is $|G| = 3$. The group element π_{120}, for example, is a rotation of the equilateral triangle by 120 degrees into itself. Using permutation notation (both Cauchy's notation and cycle notation), the three permutations in G are given as follows:

$$\pi_0 = \begin{pmatrix} 1 & 2 & 3 \\ 1 & 2 & 3 \end{pmatrix} = (1)(2)(3) = x_1 x_1 x_1 = x_1^3$$

$$\pi_{120} = \begin{pmatrix} 1 & 2 & 3 \\ 2 & 3 & 1 \end{pmatrix} = (123) = x_3$$

$$\pi_{240} = \begin{pmatrix} 1 & 2 & 3 \\ 3 & 1 & 2 \end{pmatrix} = (132) = x_3$$

From these permutations, we can now build the *cycle index*: $P_G(x_1, x_2, x_3) = \dfrac{1}{|G|} \sum_{\pi \in G} x_1^{j_1(\pi)} x_2^{j_2(\pi)} x_3^{j_3(\pi)} = \dfrac{1}{3}(x_1^3 + 2x_3)$. Now we are ready to use Polya's theorem, Theorem 14.1. For each x_j in the cycle index, let $x_j = b^j + w^j$. Note that we are using two colors of beads in this example, namely black (*b*) and white (*w*). With this substitution, the cycle index becomes

$$P_G = \frac{1}{3}\left((b^1 + w^1)^3 + 2(b^3 + w^3)\right)$$
$$= b^3 + b^2w + bw^2 + w^3.$$

This pattern inventory tells us how many ways we can form a 3-bead necklace with black (b) and white (w) beads. For example, the coefficient of b^2w in P_G is 1. So there is one way to color a 3-bead necklace using precisely two black beads and one white bead (i.e., $1b^2w$) taking only rotational symmetries into account (not flips). This necklace is shown in Figure 14.1.

The Ferris Wheel Problem

We now want to use the number theoretic method to generalize Polya's enumeration theorem so that we can count nonequivalent *object distributions* taking group symmetries into account. Consider the following generalization of Example 14.2 which we will call the "Ferris Wheel Problem." Here is the basic idea: The necklace problem involved counting the number of ways that we could assign colored beads to the vertices of a regular n-gon taking symmetries into account. For the Ferris Wheel Problem, we want to count the number of ways that we can assign objects, such as people or colored balls, into the "buckets" of a Ferris Wheel, again taking symmetries into account. What makes the problem interesting is three things: (1) We want to account for symmetries, like rotations or flips of the Ferris Wheel, and (2) it is possible that the objects, like colored balls, can occur with repetitions. We might, for example, want to count the number of ways to distribute five red balls, three white balls, and six blue balls into the buckets of the Ferris Wheel. Balls of the same color are indistinguishable. Finally, (3) Each bucket of the Ferris Wheel can hold more than one object.

Example 14.3 (Ferris Wheel Problem). Suppose we have a multiset of objects, such as colored balls, given by $S = \{r, r, b, b\}$. Here, our multiset contains two red balls and two blue balls. Now imagine that we have a triangular Ferris Wheel with three identical buckets, or boxes, that can hold objects from the set S. In how many *rotationally nonequivalent* ways can we distribute all of the objects from S into the Ferris Wheel boxes with no box empty? This problem is simple enough that we don't need a mathematical theory. Here is the answer:

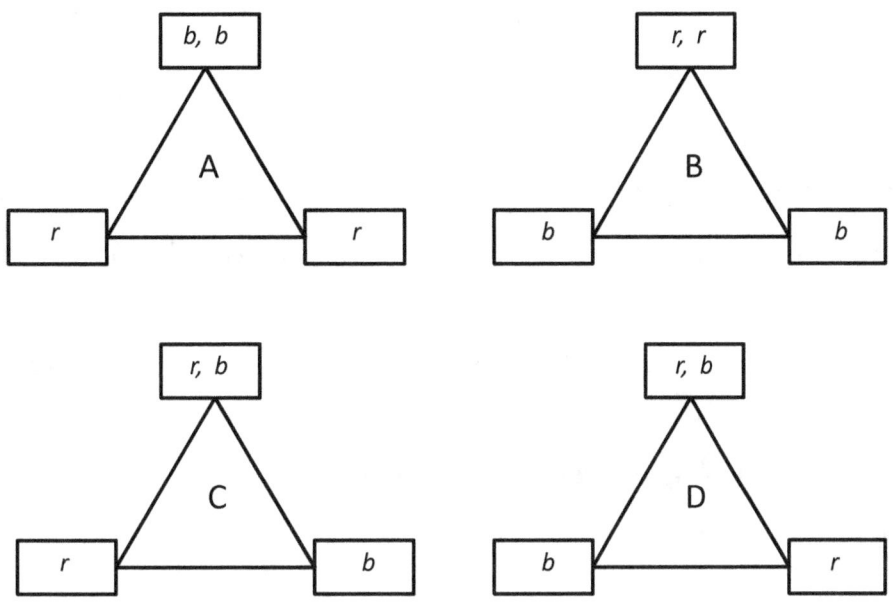

Figure 14.2. Four rotationally inequivalent Ferris Wheel distributions.

There are four fundamentally distinct Ferris Wheel distributions of the objects of S taking only *rotational* symmetries into account. (Note that configuration C cannot be rotated into configuration D.) The order of the objects within each box does not matter.

Like the necklace example in Example 14.2, we are only considering rotational symmetries, but our generalization of Polya's theorem will easily handle any symmetries, including flips.

First, we will state my generalization of Polya's theorem. Then we will give an illustrative example of the generalized theorem. Finally, we will prove the theorem. Given Polya's theorem, we will show how it can be modified, using the number theoretic method, to solve more general object distribution problems.

Generalization of Polya's Enumeration Theorem

Theorem 14.4 (Number-Theoretic Generalization of Polya's Theorem). *Let S be a set of objects, such as colored balls, having specification number n. Let X be a set of elements and G a group of permutations of X that acts to induce an equivalence relation on the labeling of X. The inventory of nonequivalent distributions of objects from S into boxes labeled by X, with no box empty, is given by*

$$P_G(x_1, x_2, \ldots, x_k) = P_G\left(\sum_{\substack{d|n \\ d \neq 1}} t\, x^{\log d}, \sum_{\substack{d|n \\ d \neq 1}} t^2 x^{2\log d}, \ldots, \sum_{\substack{d|n \\ d \neq 1}} t^k x^{k\log d}\right),$$

where P_G is the cycle index of the group G acting on X. In particular, the coefficient of $t^k x^{\log n}$ in P_G counts the number of nonequivalent object distributions of S into precisely k boxes with no box empty.

In our generalization of Polya's theorem, the order of objects within each box does not matter.

Before we prove Theorem 14.4, let's revisit the Ferris Wheel problem, Example 14.3, and solve it using Theorem 14.4.

Example 14.5. Suppose we have a multiset of objects, such as colored balls, given by $S = \{r, r, b, b\}$. In how many *rotationally nonequivalent* ways can we distribute all of the objects from S into the buckets of a Ferris Wheel shaped like an equilateral triangle with no box empty? First, we need a specification number to encode the set of objects in S. If we represent a red ball by the prime number 2, and a blue ball by the prime number 3, then we can write the set S as $S = \{2, 2, 3, 3\}$. Then the specification number for the set of objects in S is $n = 2 \cdot 2 \cdot 3 \cdot 3 = 2^2 \cdot 3^2 = 36$. The set X is $X = \{1, 2, 3\}$ viewed as the labeling set of the vertices of an equilateral triangle as shown in Figure 14.1. The group G is the group of rotations of the equilateral triangle. So, G is a set of permutations of the labeled vertices of the triangle: $G = \{\pi_0, \pi_{120}, \pi_{240}\}$. The cycle index was given in Example 14.2: $P_G = \frac{1}{3}(x_1^3 + 2x_3)$. Now, using the generalization of Polya's Enumeration theorem, Theorem 14.4, we let $x_j = \sum_{\substack{d|n \\ d \neq 1}} t^j x^{j \log d}$. Our cycle index then becomes:

$$P_G = \frac{1}{3}(x_1^3 + 2x_3)$$

$$= \frac{1}{3}\left(\left(\sum_{\substack{d|36 \\ d \neq 1}} t^1 x^{\log d}\right)^3 + 2\left(\sum_{\substack{d|36 \\ d \neq 1}} t^3 x^{3\log d}\right)\right).$$

The non-unit divisors, d, of 36 are $\{2, 3, 4, 6, 9, 12, 18, 36\}$. So, we have

$$P_G = \frac{1}{3}\left(\left(tx^{\log 2} + tx^{\log 3} + \ldots + tx^{\log 36}\right)^3 + 2\left(t^3 x^{3\log 2} + t^3 x^{3\log 3} + \ldots + t^3 x^{3\log 36}\right)\right).$$

When we algebraically expand this expression (at least, in principle) for P_G, we can pick out the terms that give rise to $t^3 x^{\log 36}$. The t^3 means that we have three boxes, or buckets, in the Ferris Wheel. The $x^{\log 36}$ means that we have a set of objects with specification number $n = 36$, which is just the set of objects in S. Although performing the complete expansion of P_G would be quite tedious, it is usually not necessary. With a little thought, we can examine the expression for P_G and identify the products that give rise to $t^3 x^{\log 36}$. Note, for example, that we can ignore all of the terms in the second part of the expression for P_G (i.e., terms arising from $2\left(t^3 x^{3\log 2} + t^3 x^{3\log 3} + \ldots + t^3 x^{3\log 36}\right)$) because all of these terms contain $x^{3\log 36}$. We want terms with $\log 36$, not $3\log 36$. Pulling out only the terms from P_G that give rise to $x^{\log 36}$, after cubing the first part of P_G, we get

$$\frac{1}{3}\left(3 tx^{\log 2} \cdot tx^{\log 9} \cdot tx^{\log 2} + 3 tx^{\log 3} \cdot tx^{\log 3} \cdot tx^{\log 4} + 6 tx^{\log 2} \cdot tx^{\log 3} \cdot tx^{\log 6}\right) = \frac{1}{3}(12 t^3 x^{\log 36})$$
$$= 4 t^3 x^{\log 36}.$$

Since the coefficient of $t^3 x^{\log 36}$ is 4, there are precisely 4 rotationally inequivalent distributions of objects from S into three Ferris Wheel buckets. This is the same answer that we obtained experimentally in Example 14.3 and illustrated in Figure 14.2.

Theorem 14.4 is entirely general. It applies to any multiset of objects, and any structure X that is acted upon by an automorphism group of symmetries G. The Ferris Wheel problem is only one special example of Theorem 14.4. Now we will prove Theorem 14.4 before giving another example of its application. The proof of Theorem 14.4 essentially consists of making a simple, but clever, modification to the existing form of Polya's theorem in order to convert it into a number-theoretic form.

Proof of Theorem 14.4. Polya's counting theorem, Theorem 14.1, uses a set of colors $C = \{c_1, c_2, \ldots, c_k\}$ as a labeling set. A necklace, for example, may have colored beads whose colors are chosen from the coloring set C. Now suppose that we have a multiset of objects, S. The objects in S may, for example, be colored balls with color repetitions allowed. According to the number theoretic paradigm, we can represent distinct objects in S by distinct prime numbers. For example, if $S = \{red, red, blue, blue, white\}$, we can let $red = 2$, $blue = 3$, and $white = 5$. Then we have $S = \{2, 2, 3, 3, 5\}$. The entire set of objects in S can be represented by a single specification number $n = 2 \cdot 2 \cdot 3 \cdot 3 \cdot 5 = 180$. The Fundamental Theorem of Arithmetic allows us to recover the primes by uniquely factoring the specification number. When we distribute objects from S into identical boxes on some structure, as was done in the Ferris Wheel problem, each box will contain a subset of elements from S. If we multiply the prime numbers within each box, then each box has its own specification number d, which must be a divisor of n, the specification number of S. In order to use Polya's counting theorem, we want to think of each box, with its set of objects, as a "color" in some coloring set C. Since the specification number, d, for each box of objects is a divisor of n, the coloring set C must be the set of all non-unit divisors of n. For example, if $S = \{red, red, blue, blue\} = \{2, 2, 3, 3\}$, then the specification number for S is $n = 2 \cdot 2 \cdot 3 \cdot 3 = 36$. The non-unit divisors of 36 are $\{2, 3, 4, 6, 9, 12, 18, 16\}$. So, the coloring set is $C = \{2, 3, 4, 6, 9, 12, 18, 36\}$. We can almost use this coloring set in Polya's counting theorem, but there are a couple problems we must circumvent. First, the coloring set, C, used in Polya's theorem is a set of variables, like b for "black" and w for "white." Our coloring set C is a set of numbers, not variables. Second, we want to know how many ways we can distribute all of the objects from S into boxes, or cells, so we need to keep track of when all the colors used for a coloring comprise the entire set of available objects in S. This would be the case if the product of all the specification number divisors, d_j, appearing in the boxes, is equal to the specification number, n, of S. We can satisfy these requirements by using a clever trick from number theoretic combinatorics. If d_j is a divisor of n, then we let the "color" c_j be $c_j = t \cdot x^{\log d_j}$. The variable t is a "box counter" that simply keeps track of the number of boxes. The expression $t \cdot x^{\log d_j}$ means that we have a specification of objects, d_j, from S distributed into one box, since the exponent of t is one. Since the logarithm function is a homomorphism, that is $\log(ab) = \log a + \log b$, the expression $t \cdot x^{\log d_j}$ automatically keeps track of products and factors when we multiply these expressions. For example, $t \cdot x^{\log(2 \cdot 3)}$ is a box with two objects $\{2, 3\}$, and $t \cdot x^{\log 3}$ is a single box with one object $\{3\}$. If we multiply these two functions, we get $t \cdot x^{\log(2 \cdot 3)} \cdot t \cdot x^{\log 3} = t^2 x^{\log(2 \cdot 3 \cdot 3)} = t^2 x^{\log 18}$. The t^2 term tells us that we have 2 boxes.

Now, since each color c_j is given by $c_j = t \cdot x^{\log d_j}$, for d_j a non-unit divisor of n, the term $\sum_{j=1}^{m} c_j^r$ in Polya's counting theorem is replaced by $\sum_{\substack{d \mid n \\ d \neq 1}} t^r x^{r \log d}$. After making this substitution in Polya's theorem, we simply expand the generating function and look for the coefficient of $x^{\log n}$. That term will give us the pattern inventory for distributing objects from S into the boxes with no box empty. □

Example 14.6. Suppose that, instead of a Ferris Wheel, we have an isosceles triangle with empty boxes at each of its vertices:

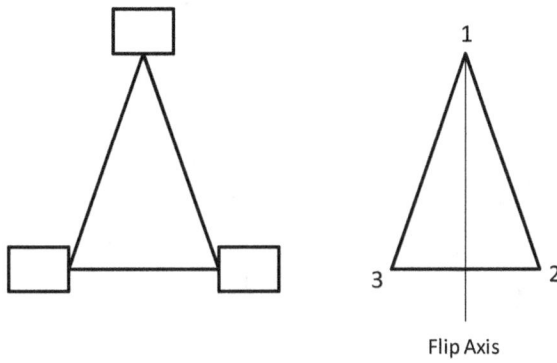

Figure 14.3. Isosceles triangle with empty boxes at each vertex.

Suppose also that we have a multiset consisting of two red balls and two blue balls: $S = \{r, r, b, b\}$. In how many nonequivalent ways can we distribute this set of colored balls into the boxes of the isosceles triangle, shown in Figure 14.3, with no box empty and taking the various symmetries of the triangle into account? We will solve this enumeration problem using the generalized Polya counting theorem, Theorem 14.4. First, let us determine a specification number, n, for the set S of objects. If we let a red ball be represented by the prime number 2, and a blue ball by the prime number 3, then S becomes $S = \{2, 2, 3, 3\}$. Multiplying together the primes in S, we obtain a specification number for the set of objects: $n = 2 \cdot 2 \cdot 3 \cdot 3 = 2^2 \cdot 3^2 = 36$. The non-unit divisors of 36 are $\{2, 3, 4, 6, 9, 12, 18, 36\}$. These are the divisors that we will use in Theorem 14.4. Next, we need to determine the symmetry group, G, for the isosceles triangle shown in the right-hand side of Figure 14.3. There are only two symmetries—the identity (which leaves the figure unchanged) and a "flip" about the flip axis. So, the group G of symmetry operations acting on the isosceles triangle consists of the following two permutations:

$$\text{Identity} = \begin{pmatrix} 1 & 2 & 3 \\ 1 & 2 & 3 \end{pmatrix} = (1)(2)(3) = x_1^3$$

$$\text{Flip} = \begin{pmatrix} 1 & 2 & 3 \\ 1 & 3 & 2 \end{pmatrix} = (1)(23) = x_1 x_2$$

The symmetry group G contains two elements (symmetry operations), so the order of G is 2: $|G| = 2$. The cycle index for the group of permutations operating on the vertices of the isosceles triangle is $P_G = \frac{1}{2}(x_1^3 + x_1 x_2)$. Now, to use Theorem 14.4, we let $x_j = \sum_{\substack{d \mid 36 \\ d \neq 1}} t^j x^{j \log d}$. Making this substitution into the cycle index, we get the following:

$$P_G = \frac{1}{2}\left(\left(\sum_{\substack{d \mid 36 \\ d \neq 1}} t^1 x^{1 \log d} \right)^3 + \left(\sum_{\substack{d \mid 36 \\ d \neq 1}} t^1 x^{1 \log d} \right)\left(\sum_{\substack{d \mid 36 \\ d \neq 1}} t^2 x^{2 \log d} \right) \right)$$

$$= \frac{1}{2}\left(\left(tx^{\log 2} + tx^{\log 3} + \ldots + tx^{\log 36} \right)^3 + (tx^{\log 2} + tx^{\log 3} + \ldots + tx^{\log 36})(t^2 x^{2\log 2} + t^2 x^{2\log 3} + \ldots + t^2 x^{2\log 36}) \right)$$

$$= \frac{1}{2}\left(\left(tx^{\log 2} + tx^{\log 3} + \ldots + tx^{\log 36} \right)^3 + (tx^{\log 2} + tx^{\log 3} + \ldots + tx^{\log 36})(t^2 x^{\log 4} + t^2 x^{\log 9} + \ldots + t^2 x^{\log 1296}) \right)$$

At this point, the solution is now entirely mechanical. We simply expand the cycle index P_G and read off the coefficient of $t^3 x^{\log 36}$. The t^3 tells us that we want three boxes, and the $x^{\log 36}$ is for the specification number, $n = 36$, for our set S of objects. In actual practice, we rarely perform the full expansion, or multiplication, of the cycle index P_G. Instead, realizing that we only want the terms that contribute to $t^3 x^{\log 36}$, we can eliminate higher order terms and make other simplifications to the analysis. We will skip the details, but you will find that we get $\frac{1}{2} \cdot 14t^3 x^{\log 36} = 7t^3 x^{\log 36}$. Don't forget that we have a "1/2" at the front end of the cycle index P_G. Since the coefficient of $t^3 x^{\log 36}$ is 7, there are 7 nonequivalent ways to distribute two red balls and two blue balls into the boxes on the isosceles triangle, with no box empty, taking all the symmetries of the triangle into account. Here are the 7 distributions:

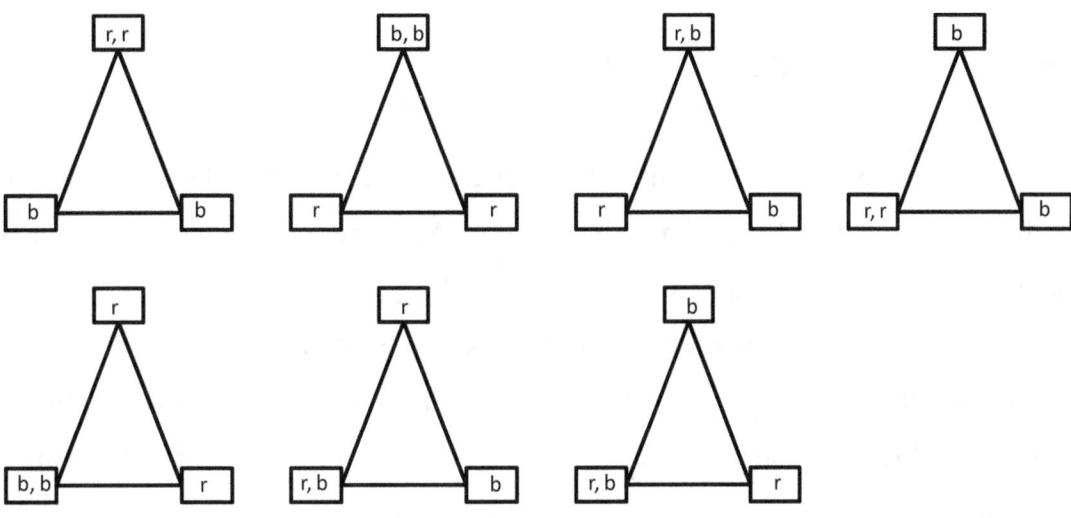

Figure 14.4. The 7 nonequivalent distributions for Example 14.6.

Let us do one more example to complete this chapter. The examples should be enough to convince you of the great power and generality of Theorem 14.4.

Example 14.7. Consider the graph shown in Figure 14.5. Suppose that we have a multiset consisting of two red balls and two blue balls: $S = \{r, r, b, b\}$. In the number-theoretic method, we need a specification number, n, that represents this set of objects. Let us represent a red ball by the prime number 2 and a blue ball by the prime number 3. Then the set S can be represented as $S = \{2, 2, 3, 3\}$. The specification number for S is $n = 2 \cdot 2 \cdot 3 \cdot 3 = 36$. There is nothing special about the primes 2 and 3. We could just as well use other prime numbers to represent the red and blue balls. We would, of course, get a different specification number for S, but all specification numbers having the same form—as powers of distinct primes—will give the same result for our final answer. Note, for example, that $2^2 \cdot 3^2$ has the same form as, say, $5^2 \cdot 13^2$, or $p^2 \cdot q^2$ for distinct primes p and q.

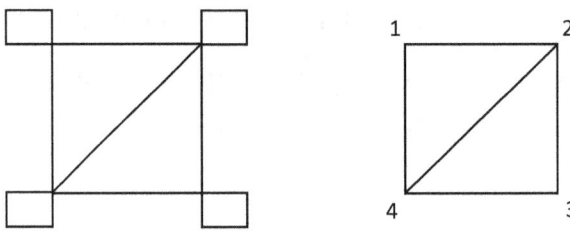

Figure 14.5. Graph for Example 14.7.

We would like to count the number of ways that we can distribute two red balls and two blue balls into the boxes in the graph of Figure 14.5, with no box empty. Also, the order of the balls within each box does not matter. However, we do want to take into account any symmetries of the graph. These symmetries will generally include rotations and flips. We represent the symmetries of the graph by the group G of permutations that act on the labeled vertices of the graph. Once we identify the group of symmetries, we can build the cycle index P_G. For the graph shown in Figure 14.5, there are four symmetries: (1) the identity, ι, which does nothing to the graph, (2) a flip, f_1, about the axis 2-4, (3) a flip, f_2, about the axis 1-3, and (4) a rotation, r, of 180 degrees. Here are the permutations of G that generate these symmetries:

$$\iota = \begin{pmatrix} 1 & 2 & 3 & 4 \\ 1 & 2 & 3 & 4 \end{pmatrix} = (1)(2)(3)(4) = x_1 x_1 x_1 x_1 = x_1^4$$

$$f_1 = \begin{pmatrix} 1 & 2 & 3 & 4 \\ 3 & 2 & 1 & 4 \end{pmatrix} = (13)(2)(4) = x_2 x_1 x_1 = x_1^2 x_2$$

$$f_2 = \begin{pmatrix} 1 & 2 & 3 & 4 \\ 1 & 4 & 3 & 2 \end{pmatrix} = (24)(1)(3) = x_2 x_1 x_1 = x_1^2 x_2$$

$$r = \begin{pmatrix} 1 & 2 & 3 & 4 \\ 3 & 4 & 1 & 2 \end{pmatrix} = (13)(24) = x_2 x_2 = x_2^2$$

The cycle index is $P_G = \frac{1}{4}\left(x_1^4 + 2x_1^2 x_2 + x_2^2\right)$. Note that we have a coefficient of 1/4 in P_G because the order of the group G is 4. That is, G has 4 elements (ι, f_1, f_2, r). Now that we know the cycle index, we can use Theorem 14.4 to solve the problem. We have

$$P_G = \frac{1}{4}\left(\left(\sum_{\substack{d|36\\d\neq 1}} tx^{\log d}\right)^4 + 2\left(\sum_{\substack{d|36\\d\neq 1}} tx^{\log d}\right)^2 \left(\sum_{\substack{d|36\\d\neq 1}} t^2 x^{2\log d}\right) + \left(\sum_{\substack{d|36\\d\neq 1}} t^2 x^{2\log d}\right)^2\right).$$

At this point, the solution is purely mechanical. We simply expand the expression for P_G, collect like terms, and read off the term containing $t^4 x^{\log 36}$. The t^4 term tells us that we have 4 boxes, and the $x^{\log 36}$ term tells us that our object set has specification number 36. As mentioned in previous examples, we don't usually need to actually expand P_G in practice. Since we know that we want the terms of the form $t^4 x^{\log 36}$, we can use some intelligent thought to determine how those terms might arise when we do expand P_G. You may have to practice to develop your skill. For this problem, expanding P_G and identifying only those terms that contribute to $t^4 x^{\log 36}$, we get

$$\frac{1}{4}\left(6t^4 x^{\log 36} + 2(2t^4 x^{\log 36}) + 2t^4 x^{\log 36}\right) = \frac{1}{4}\left(6t^4 x^{\log 36} + 4t^4 x^{\log 36} + 2t^4 x^{\log 36}\right)$$
$$= \frac{1}{4}(12t^4 x^{\log 36})$$
$$= 3t^4 x^{\log 36}.$$

Since the coefficient of $t^4 x^{\log 36}$ is 3, there are 3 inequivalent ways to distribute two red balls and two blue balls into the boxes on the graph:

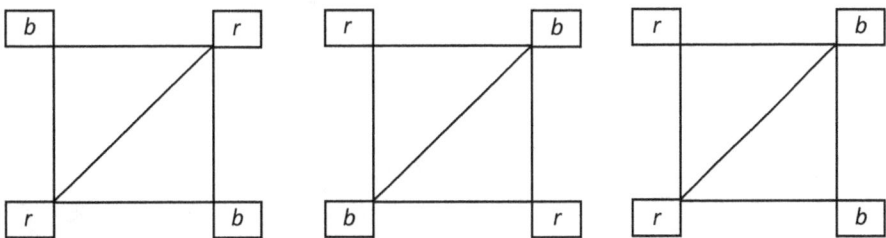

Figure 14.6. The three inequivalent distributions for Example 14.7.

Exercises

1. Calculate the cycle index for a regular pentagon taking into account both rotations and flips.
2. What is the cycle index for Problem 1 if we only account for rotations, not flips?

3. Calculate the cycle index for Example 14.5 by considering both rotations and flips of the equilateral triangle.

Appendix A. Object Distribution Polynomials

The following tables provide lists of object distribution polynomials. An object distribution polynomial, $g_n(t)$, enumerates, or counts, the number of ways to distribute objects of specification n into identical, or indistinguishable, boxes with no box empty. To read the tables, first form a specification number for your set of objects, such as colored balls. For example, if you have a multiset S consisting of three red (r) balls, two white (w) balls, and one blue (b) ball, we have $S = \{r, r, r, w, w, b\}$. Next, we assign a distinct prime number, p_j, to each color, say $p_1 = r$, $p_2 = w$, and $p_3 = b$. Then S becomes $S = \{p_1, p_1, p_1, p_2, p_2, p_3\}$. It doesn't matter which primes we use so long as a distinct prime is assigned to each color, or type, or object. Next, we multiply the primes in S to obtain a specification number that encodes the set of objects: $n = p_1^3 p_2^2 p_3$. Now we look up the corresponding polynomial $g_n(t)$ in the appropriate table. In this example, we have objects of three colors, so we use the "Trichromatic" table, Table 3. The polynomial corresponding to $n = p_1^3 p_2^2 p_3$ is $g_n(t) = t + 11t^2 + 20t^3 + 14t^4 + 5t^5 + t^6$. The coefficient of t^k in $g_n(t)$ is the number of ways to distribute three red balls, two white balls and one blue ball into precisely k identical boxes with no box empty. Here, the coefficient of t^3 is 20, for example, so there are 20 ways to distribute the colored balls into precisely three identical boxes with no box empty.

You might wonder how these polynomials were calculated. The polynomials in these tables were calculated by the author using a combination of mathematical theory and computer programs. First, the theorems and recurrence relations developed in this book were used to create algorithms. Second, the algorithms were coded as computer programs using both MATLAB® and Mathematica® software. Then the computer programs performed the actual calculations on a laptop PC. The tables are necessarily incomplete, because there are infinitely many possible object distribution polynomials.

Table 0. General Occupancy: Distinct Objects
(Stirling Numbers of the Second Kind)

Specification Number, n	$g_n(t)$
p_1	t
$p_1 p_2$	$t + t^2$
$p_1 p_2 p_3$	$t + 3t^2 + t^3$
$p_1 p_2 p_3 p_4$	$t + 7t^2 + 6t^3 + t^4$
$p_1 p_2 p_3 p_4 p_5$	$t + 15t^2 + 25t^3 + 10t^4 + t^5$
$p_1 p_2 p_3 p_4 p_5 p_6$	$t + 31t^2 + 90t^3 + 65t^4 + 15t^5 + t^6$
$p_1 p_2 p_3 p_4 p_5 p_6 p_7$	$t + 63t^2 + 301t^3 + 350t^4 + 140t^5 + 21t^6 + t^7$
$p_1 p_2 p_3 p_4 p_5 p_6 p_7 p_8$	$t + 127t^2 + 966t^3 + 1701t^4 + 1050t^5 + 266t^6 + 28t^7 + t^8$
$p_1 p_2 p_3 p_4 p_5 p_6 p_7 p_8 p_9$	$t + 255t^2 + 3025t^3 + 7770t^4 + 6951t^5 + 2646t^6 + 462t^7 + 36t^8 + t^9$
$p_1 p_2 p_3 p_4 p_5 p_6 p_7 p_8 p_9 p_{10}$	$t + 511t^2 + 9330t^3 + 34105t^4 + 42525t^5 + 22827t^6 + 5880t^7 + 750t^8 + 45t^9 + t^{10}$

Table 1. General Occupancy: Identical Objects
(Integer Partitions)

Specification Number, n	$g_n(t)$
p_1	t
p_1^2	$t + t^2$
p_1^3	$t + t^2 + t^3$
p_1^4	$t + 2t^2 + t^3 + t^4$
p_1^5	$t + 2t^2 + 2t^3 + t^4 + t^5$
p_1^6	$t + 3t^2 + 3t^3 + 2t^4 + t^5 + t^6$
p_1^7	$t + 3t^2 + 4t^3 + 3t^4 + 2t^5 + t^6 + t^7$
p_1^8	$t + 4t^2 + 5t^3 + 5t^4 + 3t^5 + 2t^6 + t^7 + t^8$
p_1^9	$t + 4t^2 + 7t^3 + 6t^4 + 5t^5 + 3t^6 + 2t^7 + t^8 + t^9$
p_1^{10}	$t + 5t^2 + 8t^3 + 9t^4 + 7t^5 + 5t^6 + 3t^7 + 2t^8 + t^9 + t^{10}$
p_1^{11}	$t + 5t^2 + 10t^3 + 11t^4 + 10t^5 + 7t^6 + 5t^7 + 3t^8 + 2t^9 + t^{10} + t^{11}$
p_1^{12}	$t + 6t^2 + 12t^3 + 15t^4 + 13t^5 + 11t^6 + 7t^7 + 5t^8 + 3t^9 + 2t^{10} + t^{11} + t^{12}$

Table 2. General Occupancy: Dichromatic Objects
(balls of two colors)

Specification Number, n	$g_n(t)$
$p_1 p_2$	$t + t^2$
$p_1^2 p_2$	$t + 2t^2 + t^3$
$p_1^2 p_2^2$	$t + 4t^2 + 3t^3 + t^4$
$p_1^3 p_2$	$t + 3t^2 + 2t^3 + t^4$
$p_1^3 p_2^2$	$t + 5t^2 + 6t^3 + 3t^4 + t^5$
$p_1^3 p_2^3$	$t + 7t^2 + 11t^3 + 8t^4 + 3t^5 + t^6$
$p_1^4 p_2$	$t + 4t^2 + 4t^3 + 2t^4 + t^5$
$p_1^4 p_2^2$	$t + 7t^2 + 10t^3 + 7t^4 + 3t^5 + t^6$
$p_1^4 p_2^3$	$t + 9t^2 + 18t^3 + 16t^4 + 9t^5 + 3t^6 + t^7$
$p_1^4 p_2^4$	$t + 12t^2 + 29t^3 + 32t^4 + 21t^5 + 10t^6 + 3t^7 + t^8$
$p_1^5 p_2$	$t + 5t^2 + 6t^3 + 4t^4 + 2t^5 + t^6$
$p_1^5 p_2^2$	$t + 8t^2 + 15t^3 + 12t^4 + 7t^5 + 3t^6 + t^7$
$p_1^5 p_2^3$	$t + 11t^2 + 26t^3 + 28t^4 + 18t^5 + 9t^6 + 3t^7 + t^8$
$p_1^5 p_2^4$	$t + 14t^2 + 42t^3 + 53t^4 + 42t^5 + 23t^6 + 10t^7 + 3t^8 + t^9$
$p_1^5 p_2^5$	$t + 17t^2 + 60t^3 + 90t^4 + 80t^5 + 52t^6 + 25t^7 + 10t^8 + 3t^9 + t^{10}$
$p_1^6 p_2$	$t + 6t^2 + 9t^3 + 7t^4 + 4t^5 + 2t^6 + t^7$
$p_1^6 p_2^2$	$t + 10t^2 + 21t^3 + 21t^4 + 13t^5 + 7t^6 + 3t^7 + t^8$
$p_1^6 p_2^3$	$t + 13t^2 + 37t^3 + 45t^4 + 34t^5 + 19t^6 + 9t^7 + 3t^8 + t^9$
$p_1^6 p_2^4$	$t + 17t^2 + 58t^3 + 86t^4 + 75t^5 + 48t^6 + 24t^7 + 10t^8 + 3t^9 + t^{10}$
$p_1^6 p_2^5$	$t + 20t^2 + 83t^3 + 142t^4 + 143t^5 + 102t^6 + 58t^7 + 26t^8 + 10t^9 + 3t^{10} + t^{11}$
$p_1^6 p_2^6$	$t + 24t^2 + 114t^3 + 224t^4 + 251t^5 + 200t^6 + 124t^7 + 64t^8 + 27t^9 + 10t^{10} + 3t^{11} + t^{12}$

Table 3. General Occupancy: Trichromatic Objects
(balls of three colors)

Specification Number, n	$g_n(t)$
$p_1 p_2 p_3$	$t + 3t^2 + t^3$
$p_1^2 p_2 p_3$	$t + 5t^2 + 4t^3 + t^4$
$p_1^2 p_2^2 p_3$	$t + 8t^2 + 11t^3 + 5t^4 + t^5$
$p_1^2 p_2^2 p_3^2$	$t + 13t^2 + 26t^3 + 19t^4 + 6t^5 + t^6$
$p_1^3 p_2 p_3$	$t + 7t^2 + 8t^3 + 4t^4 + t^5$
$p_1^3 p_2^2 p_3$	$t + 11t^2 + 20t^3 + 14t^4 + 5t^5 + t^6$
$p_1^3 p_2^2 p_3^2$	$t + 17t^2 + 46t^3 + 44t^4 + 22t^5 + 6t^6 + t^7$
$p_1^3 p_2^3 p_3$	$t + 15t^2 + 36t^3 + 34t^4 + 17t^5 + 5t^6 + t^7$
$p_1^3 p_2^3 p_3^2$	$t + 23t^2 + 80t^3 + 100t^4 + 64t^5 + 25t^6 + 6t^7 + t^8$
$p_1^3 p_2^3 p_3^3$	$t + 31t^2 + 139t^3 + 219t^4 + 175t^5 + 86t^6 + 28t^7 + 6t^8 + t^9$
$p_1^4 p_2 p_3$	$t + 9t^2 + 14t^3 + 9t^4 + 4t^5 + t^6$
$p_1^4 p_2^2 p_3$	$t + 14t^2 + 33t^3 + 29t^4 + 15t^5 + 5t^6 + t^7$
$p_1^4 p_2^2 p_3^2$	$t + 22t^2 + 73t^3 + 88t^4 + 55t^5 + 23t^6 + 6t^7 + t^8$
$p_1^4 p_2^3 p_3$	$t + 19t^2 + 58t^3 + 67t^4 + 43t^5 + 18t^6 + 5t^7 + t^8$
$p_1^4 p_2^3 p_3^2$	$t + 29t^2 + 126t^3 + 191t^4 + 151t^5 + 75t^6 + 26t^7 + 6t^8 + t^9$
$p_1^4 p_2^3 p_3^3$	$t + 39t^2 + 216t^3 + 411t^4 + 394t^5 + 235t^6 + 97t^7 + 29t^8 + 6t^9 + t^{10}$
$p_1^4 p_2^4 p_3$	$t + 24t^2 + 92t^3 + 129t^4 + 101t^5 + 52t^6 + 19t^7 + 5t^8 + t^9$
$p_1^4 p_2^4 p_3^2$	$t + 37t^2 + 196t^3 + 361t^4 + 339t^5 + 205t^6 + 86t^7 + 27t^8 + 6t^9 + t^{10}$
$p_1^4 p_2^4 p_3^3$	$t + 49t^2 + 334t^3 + 761t^4 + 864t^5 + 606t^6 + 298t^7 + 108t^8 + 30t^9 + 6t^{10} + t^{11}$
$p_1^4 p_2^4 p_3^4$	$t + 62t^2 + 513t^3 + 1399t^4 + 1857t^5 + 1513t^6 + 855t^7 + 364t^8 + 119t^9 + 31t^{10} + 6t^{11} + t^{12}$

Table 4. General Occupancy: Tetrachromatic Objects
(balls of four colors)

Specification Number, n	$g_n(t)$
$p_1 p_2 p_3 p_4$	$t + 7t^2 + 6t^3 + t^4$
$p_1^2 p_2 p_3 p_4$	$t + 11t^2 + 16t^3 + 7t^4 + t^5$
$p_1^3 p_2 p_3 p_4$	$t + 15t^2 + 30t^3 + 20t^4 + 7t^5 + t^6$
$p_1^2 p_2^2 p_3 p_4$	$t + 17t^2 + 38t^3 + 27t^4 + 8t^5 + t^6$
$p_1^4 p_2 p_3 p_4$	$t + 19t^2 + 49t^3 + 43t^4 + 21t^5 + 7t^6 + t^7$
$p_1^3 p_2^2 p_3 p_4$	$t + 23t^2 + 68t^3 + 66t^4 + 31t^5 + 8t^6 + t^7$
$p_1^2 p_2^2 p_3^2 p_4$	$t + 26t^2 + 85t^3 + 87t^4 + 40t^5 + 9t^6 + t^7$
$p_1^5 p_2 p_3 p_4$	$t + 23t^2 + 72t^3 + 78t^4 + 47t^5 + 21t^6 + 7t^7 + t^8$
$p_1^4 p_2^2 p_3 p_4$	$t + 29t^2 + 108t^3 + 132t^4 + 81t^5 + 32t^6 + 8t^7 + t^8$
$p_1^3 p_2^3 p_3 p_4$	$t + 31t^2 + 120t^3 + 152t^4 + 96t^5 + 35t^6 + 8t^7 + t^8$
$p_1^3 p_2^2 p_3^2 p_4$	$t + 35t^2 + 148t^3 + 198t^4 + 124t^5 + 44t^6 + 9t^7 + t^8$
$p_1^2 p_2^2 p_3^2 p_4^2$	$t + 40t^2 + 183t^3 + 259t^4 + 163t^5 + 55t^6 + 10t^7 + t^8$
$p_1^6 p_2 p_3 p_4$	$t + 27t^2 + 100t^3 + 129t^4 + 92t^5 + 48t^6 + 21t^7 + 7t^8 + t^9$
$p_1^5 p_2^2 p_3 p_4$	$t + 35t^2 + 156t^3 + 230t^4 + 171t^5 + 85t^6 + 32t^7 + 8t^8 + t^9$
$p_1^4 p_2^3 p_3 p_4$	$t + 39t^2 + 188t^3 + 294t^4 + 229t^5 + 111t^6 + 36t^7 + 8t^8 + t^9$
$p_1^4 p_2^2 p_3^2 p_4$	$t + 44t^2 + 231t^3 + 380t^4 + 298t^5 + 141t^6 + 49t^7 + 9t^8 + t^9$
$p_1^3 p_2^3 p_3^2 p_4$	$t + 47t^2 + 256t^3 + 436t^4 + 350t^5 + 163t^6 + 48t^7 + 9t^8 + t^9$
$p_1^3 p_2^2 p_3^2 p_2^2$	$t + 53t^2 + 314t^3 + 562t^4 + 459t^5 + 209t^6 + 59t^7 + 10t^8 + t^9$

Appendix B. Research Problems

There are a number of open research questions that you may wish to investigate. Several research problems are suggested here. In addition, almost any of the topics presented in the various chapters of this book can be extended or generalized in several directions.

Problem 1. Prove that the coefficients in an object distribution polynomial $g_n(t)$, taken in order by ascending powers of t, always form a unimodal sequence. A unimodal sequence increases to a maximum and then decreases.

Problem 2. Find the value of k, which may be a function of n, such that $g(n, k)$ is the maximum valued coefficient in the object distribution polynomial $g_n(t) = \sum_{k=1}^{m} g(n, k) t^k$. Often, but not always, $g_{\max}(n, k) = m/2$.

Problem 3. Find the minimum value of the real exponent r such that $g(n) \leq n^r$ for all positive integers n. Note that $g(n) = g_n(1)$ in $g_n(t)$. It is known that $r \leq 1.7265$, but numerical data show that this is clearly not the minimum value of r.

Problem 4. Prove or disprove the conjecture that $\sum_{n=1}^{\infty} \frac{\sqrt{3g(n)}}{n^2} = \pi$.

Problem 5. Find the largest positive real constant r such that $\sum_{n=1}^{m} g(n) \geq \sum_{n=1}^{m} n^r$ for all positive integers m greater than some sufficiently large positive integer N. (There exist numbers r and N such that the statement is true. We want to find the largest value of r. The author of this book proved that $0.3 < r < 0.35$.)

Problem 6. How can we determine if a given polynomial with positive integer coefficients is an object distribution polynomial for some specification of objects? Determine necessary and sufficient conditions.

A trivial necessary condition is that $g_n(t)$ must be a monic polynomial. If a polynomial $f(t)$ of degree m is given, then the specification number n for the set of objects must encode precisely m objects. Since the combinatorial possibilities for n, having m not-necessarily distinct objects, is finite, this implies that there are only a finite number of possible candidates $g_n(t)$ for $f(t)$. The partitions of m determine the possible forms of n. So, given a polynomial $f(t)$, having positive integer coefficients, we could simply compare it with a known list of object distribution polynomials $g_n(t)$ to see if there is a match. This approach is, of course, a brute force method.

Problem 7. Develop a method for enumerating all of the symmetric object distributions. A distribution is said to be *symmetric* if it is invariant under any permutation of the object type labels (i.e., object colors).

Problem 8. Find computationally efficient recurrence relations for calculating the object distribution polynomials $g_n(t)$. In particular, find a recurrence relation, if possible, that does not require summation over the divisors of the specification number n.

Problem 9. Examine the parity of the function $g(n)$. Here, $g(n) = g_n(1)$ in $g_n(t)$. For what values of n is $g(n)$ even or odd? Numerical data suggest that, on average, $g(n)$ is even as often as it is odd. That is, for randomly chosen positive integer n, Probability($g(n) =$ even) = Probability($g(n) = odd$).

Problem 10. Discover useful asymptotic relations for the behavior of object distribution polynomials.

Problem 11. Discover useful inequalities that provide good upper and lower bounds on the object distribution functions.

Appendix C. Solutions to the Exercises

Chapter 1.

1. $S(n, k)$ counts the number of ways to partition an n-set, like $\{1, 2, \ldots, n\}$, into k non-empty disjoint subsets for $1 \leq k \leq n$. To derive a recurrence relation for $S(n, k)$, consider what happens to the number 1. Either 1 is by itself in a subset, or 1 is in a subset with other elements. If 1 is by itself, then the remaining $n-1$ elements can be partitioned into $k-1$ subsets in $S(n-1, k-1)$ ways. If 1 is not by itself, then first partition the remaining $n-1$ elements into k subsets in $S(n-1, k)$ ways. Having done this, the element 1 can now be placed into any one of the k subsets, for a total of $k \cdot S(n-1, k)$ ways. Adding these two disjoint cases, we get the final result: $S(n, k) = S(n-1, k-1) + k \cdot S(n-1, k)$. With a little thought, the initial conditions are $S(n, 1) = S(n, n) = 1$.

2. A partition of m identical objects into r identical boxes with no box empty can be viewed as a representation of the number m as the sum of the numbers of objects in each of the boxes. This is just a partition of m into r parts.

3. The binomial coefficient $\binom{n}{r}$ counts the number of combinations of n distinct things chosen r at a time. For each choice of r objects, we leave behind $n-r$ objects. So there is one combination of r objects corresponding bijectively to each combination of $n-r$ objects. Hence, we have the binomial identity $\binom{n}{r} = \binom{n}{n-r}$.

4. Using the recurrence relation from exercise 1, we have
$$\begin{aligned} S(4, 2) &= S(3, 1) + 2S(3, 2) \\ &= S(3, 1) + 2(S(2, 1) + 2S(2, 2)) \\ &= 1 + 2(1 + 2(1)) \\ &= 7. \end{aligned}$$

The 7 set partitions of the set $\{1, 2, 3, 4\}$ into 2 non-empty disjoint subsets are $\{1\} \cup \{2, 3, 4\}$, $\{2\} \cup \{1, 3, 4\}$, $\{3\} \cup \{1, 2, 4\}$, $\{4\} \cup \{1, 2, 3\}$, $\{1, 2\} \cup \{3, 4\}$, $\{2, 3\} \cup \{1, 4\}$, $\{1, 3\} \cup \{2, 4\}$.

Chapter 2.

1. By Theorem 2.8, $\prod_{d=2}^{\infty} \frac{1}{(1-tx^{\log d})} = \sum_{n=1}^{\infty} g_n(t) x^{\log n}$. Since $36 = 2^2 \cdot 3^2$, the non-unit divisors of 36 are $\{2, 3, 4, 6, 9, 12, 18, 36\}$. So we only need to look at these divisors. If we expand the left-hand side of $\prod_{d \in \{2, 3, 4, 6, 9, 12, 18, 36\}} \frac{1}{(1-tx^{\log d})} = \sum_{n} g_n(t) x^{\log n}$, and collect like terms that equate to $x^{\log 36}$, we find that the coefficient of $x^{\log 36}$ is $g_{36}(t) = t + 4t^2 + 3t^3 + t^4$.

2. Since the specification number is $n = 36 = 2^2 \cdot 3^2 = 2 \cdot 2 \cdot 3 \cdot 3$, we can represent the multiset of objects as, say, $S = \{2, 2, 3, 3\} = \{red, red, blue, blue\}$. Thus, $g_{36}(t)$ gives the number of ways to distribute two red balls and two blue balls into identical boxes with no box empty. The coefficient of t^k in $g_{36}(t)$ is the number of such distributions into precisely k identical boxes with no box empty.

3. $g_{36}(t) = g_{100}(t)$ because the specification numbers 36 and 100 both have the same form when we prime factorize them: $36 = 2^2 \cdot 3^2$ and $100 = 2^2 \cdot 5^2$. Both of these specification numbers have the form $n = p_1^2 p_2^2$ for distinct primes p_1 and p_2.

4. Let $r = 2$, $w = 3$, $b = 5$, and $g = 7$ be distinct primes. Then a specification number for S is $n = 2^2 \cdot 3 \cdot 5^2 \cdot 7 = 2100$. The specification number for S is not unique. Any distinct primes could be assigned to r, w, b, and g.

5. A specification number for 3 white balls, 3 red balls, and 3 blue balls has the general form $n = p_1^3 p_2^3 p_3^3$ for distinct primes p_1, p_2, and p_3. From Appendix A, Table 3, we have $g_n(t) = t + 31t^2 + 139t^3 + 219t^4 + 175t^5 + 86t^6 + 28t^7 + 6t^8 + t^9$.

Chapter 3.

1. $\sum_{n=1}^{\infty} \frac{1}{k^{2n}}$ is a geometric progression with first term $a = \frac{1}{k^2}$ and common ratio $r = \frac{1}{k^2}$. The sum of this infinite series is $\frac{a}{1-r} = \frac{1}{k^2-1}$ for $k > 1$.

2. $h(n)$ is the number of ordered factorizations of n into non-unit factors. Thus, we have

$$h(12) = \sum_{\substack{d|12 \\ d>1}} h(12/d)$$
$$= h(12/2) + h(12/3) + h(12/4) + h(12/6) + h(12/12)$$
$$= h(6) + h(4) + h(3) + h(2) + h(1)$$
$$= 3 + 2 + 1 + 1 + 1$$
$$= 8.$$

The 8 ordered factorizations of 12 into non-unit factors are 12, $2 \cdot 6$, $6 \cdot 2$, $3 \cdot 4$, $4 \cdot 3$, $3 \cdot 2 \cdot 2$, $2 \cdot 3 \cdot 2$, and $2 \cdot 2 \cdot 3$.

Chapter 4.

1. By Corollary 4.10, $\frac{d}{dt} g_n(t)\big|_{t=1} \geq \tau(n) - 1$. But for n composite, $\tau(n) - 1 \geq \frac{1}{2}\tau(n)$, because $\tau(n) \geq 2$ for composite n.

2. Let $n = p_1^2 p_2^2 q$ for distinct primes p_1, p_2, and q. By Theorem 4.2, we have

$$g_n(t) = g_{(p_1^2 p_2^2)q}(t)$$
$$= t \sum_{d | p_1^2 p_2^2} g_d(t)$$
$$= t(g_1(t) + g_{p_1}(t) + g_{p_2}(t) + g_{p_1^2}(t) + g_{p_2^2}(t) + g_{p_1 p_2}(t) + g_{p_1^2 p_2}(t) + g_{p_1 p_2^2}(t) + g_{p_1^2 p_2^2}(t)).$$

Now look up the polynomials in Appendix A, substitute into the above expression, and simplify to get $g_n(t) = t + 8t^2 + 11t^3 + 5t^4 + t^5$.

3. By Theorem 4.7, we have, with $B_0 = 1$, $B_{k+1} = \sum_{r=0}^{k} \binom{k}{r} B_r$. We can use this recurrence formula to find $B_1 = 1$. Then, knowing B_0 and B_1, we can again use the recurrence to find $B_2 = 2$. Continuing in this way, we obtain the sequence 1, 1, 2, 5, 15, 52. So, $B_5 = 52$.

4. By Theorem 4.13, the average number of distributions per box is given by $A = \dfrac{B(k+1) - B(k)}{B(k)}$. As in problem 3, the Bell numbers form the sequence 1, 1, 2, 5, 15, 52, 203, ..., with $B_0 = B_1 = 1$. For five distinct objects, $k = 5$. Then $A = \dfrac{B(6) - B(5)}{B(5)} = \dfrac{203 - 52}{52} = 2.9$.

5. By Theorem 4.16, we have $g_n(t) = \sum_{\substack{d \mid n \\ d \neq n}} \int_0^t \dfrac{1}{u} g_d(u) P_{\frac{n}{d}}(u)\, du$. With $n = 18$, the divisors d of 18 are $\{1, 2, 3, 6, 9, 18\}$. Thus, we have $g_{18}(t) = \sum_{\substack{d \mid 18 \\ d \neq 18}} \int_0^t \dfrac{1}{u} g_d(u) P_{\frac{18}{d}}(u)\, du$. We find $P_{\frac{18}{d}}(u)$ for each divisor d using Definition 4.14 (see Example 4.15). If we perform the calculation, we get $g_{18}(t) = t + 2t^2 + t^3$.

6. (a) Since the specification number is $n = 18 = 2 \cdot 3^2$, we have, for example, one red ball (the prime 2) and two blue balls (the primes 3 and 3 in 3^2). So, $g_{18}(t)$ counts the number of ways to distribute one red ball and two blue balls into identical boxes with no box empty.

   (b) Yes, $g_{12}(t) = g_{18}(t)$. Note that $12 = 3 \cdot 2^2$ and $18 = 2 \cdot 3^2$. Both specification numbers 12 and 18 have the form $n = p_1 p_2^2$ for distinct primes p_1 and p_2.

Chapter 5.

1. By Theorem 5.11, we have $k \cdot \pi_d(k) = \sum_{j=0}^{k-1} \pi_d(j) \sigma_o(k - j)$ with $\pi_d(0) = 1$. For $k = 5$, we have
$$5 \cdot \pi_d(5) = \sum_{j=0}^{4} \pi_d(j) \sigma_o(k - j) = \pi_d(0) \sigma_o(5) + \pi_d(1) \sigma_o(4) + \pi_d(2) \sigma_o(3) + \pi_d(3) \sigma_o(2) + \pi_d(4) \sigma_o(1).$$
$\pi_d(k)$ is the number of partitions of k into distinct parts, and $\sigma_o(n)$ is the sum of the odd divisors of n. So $\pi_d(0) = 1$, $\pi_d(1) = 1$, $\pi_d(2) = 1$, $\pi_d(3) = 2$, and $\pi_d(4) = 2$. Also, $\sigma_o(5) = 6$, $\sigma_o(4) = 1$, $\sigma_o(3) = 4$, $\sigma_o(2) = 1$, and $\sigma_o(1) = 1$. Substituting these values into the equation and solving for $\pi_d(5)$, we get $\pi_d(5) = 3$. The three partitions of 5 into distinct parts are 5, 3 + 2, and 4 + 1.

2. From Table 5.1, for $n = p_1^3 p_2^3$, we have $\delta_n(t) = t + 7t^2 + 8t^3 + t^4$. For three identical boxes, we look at the coefficient fo t^3, which is 8. There are 8 possible distributions.

3. Theorem 5.16 states that $P(x) = e^{D(x)}$, where $P(x) = \sum_{n=0}^{\infty} \pi(n)x^n$ and $D(x) = \sum_{n=1}^{\infty} \frac{\sigma(n)}{n} x^n$. We have

$$P(x) = e^{D(x)} = e^{x+3x^2/2+4x^3/3+7x^4/4+6x^5/5+\ldots} = 1 + x + 2x^2 + 3x^3 + 5x^4 + \ldots .$$ Note, for example, that the coefficient of x^4 is 5, and there are, indeed, 5 partitions of the number 4, namely, 4, 3 + 1, 2 + 2, 2 + 1 + 1, and 1 + 1 + 1 + 1.

4. $\delta_{36}(t) * g_{36}(-t) = \sum_{d|36} \delta_d(t) g_{36/d}(t)$. Use the polynomials in Table 5.1 and Appendix A, substitute into this expression, and calculate the sum to get zero. Note that the divisors d of 36 are {1, 2, 3, 4, 6, 9, 12, 18, 36}.

5. By Theorem 5.29, we have $F(18) = \sum_{\substack{d|18 \\ d \neq 18}} F(d) = F(1) + F(2) + F(3) + F(6) + F(9) = 1 + 1 + 1 + 3 + 2$

= 8. The eight ordered factorizations of 18 into non-unit parts are 18, $2 \cdot 9$, $9 \cdot 2$, $3 \cdot 6$, $6 \cdot 3$, $2 \cdot 3 \cdot 3$, $3 \cdot 2 \cdot 3$, $3 \cdot 3 \cdot 2$.

Chapter 6.

1. We are given that $g(36) = \delta(36) * \varepsilon(36) = \sum_{d|36} \delta(d)\varepsilon(n/d)$. The divisors d of 36 are {1, 2, 3, 4, 6, 9, 12, 18, 36}. Note also that $g(36) = g_{36}(1)$, where $g_{36}(t) = t + 4t^2 + 3t^3 + t^4$. So, $g(36) = 1 + 4 + 3 + 1 = 9$. Similarly, $\delta(d) = \delta_d(1)$ and $\varepsilon(n/d) = \varepsilon_{n/d}(1)$. Look up polynomials in Table 5.1 and Appendix A and use the fact that $g_n(t) = \varepsilon_{n^2}(t)$. Plug the values into the right-hand side of the equation, above, and compute the sum to get 9.

2. By Theorem 6.29, we have $\pi(k) = \sum_{r=0}^{\lfloor k/2 \rfloor} \pi_e(2r) \cdot \pi_d(k - 2r)$. For $k = 6$, we have

$$\pi(6) = \sum_{r=0}^{3} \pi_e(2r)\pi_d(6-2r) = \pi_e(0)\pi_d(6) + \pi_e(2)\pi_d(4) + \pi_e(4)\pi_d(2) + \pi_e(6)\pi_d(0) =$$

$1 \cdot 4 + 1 \cdot 2 + 2 \cdot 1 + 3 \cdot 1 = 4 + 2 + 2 + 3 = 11$. There are 11 partitions of 6.

3. Since $900 = 2^2 \cdot 3^2 \cdot 5^2$, by Theorem 6.34 we have

$$g(900) \geq \sum_{r=0}^{3} \binom{3}{r} B(r) B(3-r) = \binom{3}{0} B(0) B(3) + \binom{3}{1} B(1) B(2) + \binom{3}{2} B(2) B(1) + \binom{3}{3} B(3) B(0) = 22.$$

Note that $B(n)$ is the nth Bell number, with $B(0) = 1$.

4. This problem is a straightforward calculation. The divisors d of 30 are $\{1, 2, 3, 5, 6, 10, 15, 30\}$. Look up the polynomials in Appendix A, set $t=1$ or $t=-1$, as appropriate, and calculate
$$\sum_{d|30} g_d(1) g_{\frac{30}{d}}(-1).$$

5. For $n=4$, the Square Root Theorem, Theorem 6.43, states that $g_{4^2}(t) * g_{4^2}(-t) = g_4(t^2)$. Equivalently, we have $\sum_{d|16} g_d(t) g_{\frac{16}{d}}(-t) = g_{16}(t^2)$. Look up the polynomials in Appendix A and perform the calculation on the left-hand side of the equation to get $g_{16}(t^2)$. See Example 6.44.

Chapter 7.

1. By Theorem 7.7, we have $\pi(8) = \sum_{j \equiv 0 (\mathrm{mod} \, 3)} \pi_T(j) \pi_D(8-j) = \pi_T(0) \pi_D(8) + \pi_T(3) \pi_D(5) + \pi_T(6) \pi_D(2) = 1 \cdot 13 + 1 \cdot 5 + 2 \cdot 2 = 22$. Note that $\pi_T(0) = 1$, $\pi_T(3) = 1$, $\pi_T(6) = 2$, and $\pi_D(8) = 13$, $\pi_D(5) = 5$, $\pi_D(2) = 2$. We compute the sum over non-negative integers that are divisible by 3 and less than or equal to 8.

2. Let's pick, say, $m=5$, but we could use other values of m. By Theorem 7.23, we have
$$\pi(8) = \sum_{j \equiv 0 (\mathrm{mod} \, 5)} \pi_{\mu(5)}(j) \pi_{\rho(5)}(8-j) = \pi_{\mu(5)}(0) \pi_{\rho(5)}(8) + \pi_{\mu(5)}(5) \pi_{\rho(5)}(3) = 1 \cdot 19 + 1 \cdot 3 = 22.$$
Note that $\pi_{\mu(5)}(0) = 1$, $\pi_{\mu(5)}(5) = 1$, $\pi_{\rho(5)}(8) = 19$, and $\pi_{\rho(5)}(3) = 3$. You may need to consult a list of integer partitions to determine some of these values.

3. We will demonstrate Theorem 7.25 for the case where $m=2$. By Theorem 7.25, we have
$$\pi(8) = \sum_{j \equiv 0 (\mathrm{mod} \, 2)} \pi(j \,|\, \text{all parts are even}) \pi(8-j \,|\, \text{all parts are odd}) = \pi_e(0) \pi_o(8) + \pi_e(2) \pi_o(6) + \pi_e(4) \pi_o(4) + \pi_e(6) \pi_o(2) + \pi_e(8) \pi_o(0) = 1 \cdot 6 + 1 \cdot 4 + 2 \cdot 2 + 3 \cdot 1 + 5 \cdot 1 = 22.$$

Chapter 8.

1. By Lemma 8.4, we have
$t \cdot g_{p_1 p_2}(t) = \mu(p_1 p_2) g_{p_3}(t) + \mu(p_2) g_{p_1 p_3}(t) + \mu(p_1) g_{p_2 p_3}(t) + \mu(1) g_{p_1 p_2 p_3}(t)$. This simplifies to
$t \cdot g_{p_1 p_2}(t) = (-1)^2 g_{p_3}(t) - g_{p_1 p_3}(t) - g_{p_2 p_3}(t) + g_{p_1 p_2 p_3}(t)$. Solving for $g_{p_1 p_2 p_3}(t)$, we get
$g_{p_1 p_2 p_3}(t) = t \cdot g_{p_1 p_2}(t) - g_{p_3}(t) + g_{p_1 p_3}(t) + g_{p_2 p_3}(t) = t(t+t^2) - t + (t+t^2) + (t+t^2) = t + 3t^2 + t^3$.

2. By Theorem 8.6, we have

$$g_{(p_1p_2)p_3}(t) = t \cdot g_{p_1p_2}(t) + \sum_{r=1}^{2}(-1)^{r+1}\binom{2}{r}g_{\underbrace{p_1p_2p_3}_{p_1\cdots p_r}}(t) = t \cdot g_{p_1p_2}(t) + \binom{2}{1}g_{p_2p_3}(t) - \binom{2}{2}g_{p_3}(t) =$$

$$t(t+t^2) + 2(t+t^2) - t = t + 3t^2 + t^3.$$

3. We can either calculate or look up in a reference book the binomial numbers $\binom{8}{1}=8$, $\binom{8}{2}=28$, $\binom{8}{3}=56$, $\binom{8}{4}=70$, $\binom{8}{5}=56$, and $\binom{8}{6}=28$. Likewise, the Stirling numbers of the second kind are $\left\{{3\atop 3}\right\}=1$, $\left\{{4\atop 3}\right\}=6$, $\left\{{5\atop 3}\right\}=25$, $\left\{{6\atop 3}\right\}=90$, $\left\{{7\atop 3}\right\}=301$, $\left\{{8\atop 3}\right\}=966$. Substituting these numbers into $\binom{8}{1}\left\{{8\atop 3}\right\}+\binom{8}{3}\left\{{6\atop 3}\right\}+\binom{8}{5}\left\{{4\atop 3}\right\}$, we get $8\cdot 966 + 56\cdot 90 + 56\cdot 6 = 13104$, and substituting the appropriate numbers into $\binom{8}{2}\left\{{7\atop 3}\right\}+\binom{8}{4}\left\{{5\atop 3}\right\}+\binom{8}{6}\left\{{3\atop 3}\right\}$, we get $28\cdot 301 + 70\cdot 25 + 28\cdot 1 = 10206$. Now, we note that $13104 \equiv 10206 \pmod{3}$, because both 13104 and 10206 are divisible by 3.

4. Let $p = p_1$ and $q = p_2$ be distinct primes. Also, let $m = 1$. By Theorem 8.14, we have $g_{p_1p_2}(t) = g_{p_2}(t) = t \cdot g_{p_1}(t)$, or $g_{pq}(t) = g_q(t) + t\cdot g_p(t)$. Thus, $t + t^2 = t + t\cdot t$.

Chapter 9.

1. $g_8(t) * g_8(t) = \sum_{d|8} g_d(t) g_{\frac{8}{d}}(t) = g_1(t)g_8(t) + g_2(t)g_4(t) + g_4(t)g_2(t) + g_8(t)g_1(t) = 1\cdot(t+t^2+t^3) + t\cdot(t+t^2) + (t+t^2)\cdot t + (t+t^2+t^3)\cdot 1 = 2t + 4t^2 + 4t^3$. The combinatorial meaning of this polynomial is that it counts the number of ways to distribute objects of specification $n = 8 = 2^3$, such as 3 red balls, into boxes of two colors, say black and white, with no box empty. See Example 9.4. The coefficient of t^k is the number of such distributions into precisely k boxes, where the k boxes can be of any combination, with color repetitions allowed, of two colors.

2. We will show the solution for just a few of the 11 partitions of 6 to illustrate the notation:
$6 = 3+3 = [3^2]$, $6 = 3+2+1 = [1^12^13^1]$, $6 = 4+1+1 = [1^24^1]$.

3. The 11 partitions of 6 are $[6^1]$, $[1^15^1]$, $[2^14^1]$, $[1^24^1]$, $[3^2]$, $[1^12^13^1]$, $[1^33^1]$, $[2^3]$, $[1^22^2]$, $[1^42^1]$, $[1^6]$.

Following Example 9.11, we have

$$g(6, r, t) = \binom{r}{1}t + \binom{r}{1}\binom{r}{1}t^2 + \binom{r}{1}\binom{r}{1}t^2 + \binom{r+1}{2}\binom{r}{1}t^3 + \binom{r+1}{2}t^2 +$$

$$\binom{r}{1}\binom{r}{1}\binom{r}{1}t^3 + \binom{r}{1}\binom{r+2}{3}t^4 + \binom{r+2}{3}t^3 + \binom{r+1}{2}\binom{r+1}{2}t^4 + \binom{r}{1}\binom{r+3}{4}t^5.$$ This simplifies to

$$g(6, r, t) = rt + \tfrac{1}{2}rt^2 + \tfrac{1}{3}rt^3 + \tfrac{5}{2}r^2t^2 + r^2t^3 + \tfrac{7}{12}r^2t^4 + \tfrac{1}{4}r^2t^5 + \tfrac{5}{3}r^3t^3 + r^3t^4 + \tfrac{11}{24}r^3t^5 + \tfrac{5}{12}r^4t^4 + \tfrac{1}{4}r^4t^5 + \tfrac{1}{24}r^5t^5.$$

If you want, you can collect like powers of t together. Then the coefficient of t^k will be the number of distributions into precisely k boxes, with no box empty, using r colors for the boxes. Note that the boxes can be any combination of colors, with repetitions allowed, selected from r possible colors. You could, for example, have all of the boxes be the same color.

4. For 2 red balls and 2 blue balls, let the specification number be $n = 2^2 \cdot 3^2 = 36$. From Example 9.12, we have $g_{36}(r, t) = rt + (\tfrac{1}{2}r + \tfrac{7}{2}r^2)t^2 + (r^2 + 2r^3)t^3 + (\tfrac{1}{4}r^2 + \tfrac{1}{2}r^3 + \tfrac{1}{4}r^4)t^4$. For 3 boxes, we look at the coefficient of t^3, which is $r^2 + 2r^3$. Then, for $r = 6$ different possible colors of boxes, the final answer is $6^2 + 2 \cdot 6^3 = 468$.

Chapter 10.

1. $h_6(t) * h_6(-t) = \sum_{d|6} h_d(t) h_{\frac{6}{d}}(-t) = h_1(t)h_6(-t) + h_2(t)h_3(-t) + h_3(t)h_2(-t) + h_6(t)h_1(-t) =$

   $1 \cdot (-t + t^2) + t(-t) + t(-t) + (t + t^2) \cdot 1 = 0$.

2. By Theorem 10.27, we have $f_8(t) = \tfrac{1}{8}\sum_{k=1}^{8} k \cdot t \cdot f_{8-k}(t) = \tfrac{1}{8}(t \cdot f_7(t) + 2t \cdot f_6(t) + \ldots + 8t \cdot f_0(t))$ with $f_0(t) = 1$. Now use back-substitution with the same recurrence formula for $f_r(t)$ to calculate $f_8(t)$. Another approach is to simply build up the exponential polynomials $f_r(t)$ in sequential order:

   $f_0(t) = 1$, $f_1(t) = t$, $f_2(t) = t + \dfrac{t^2}{2!}$, $f_3(t) = t + 2\dfrac{t^2}{2!} + \dfrac{t^3}{3!}$, and so on.

3. $\tau_3(36) = \tau_{2+1}(36) = \sum_{d|36} \tau_2(d) = \tau_2(1) + \tau_2(2) + \tau_2(3) + \tau_2(4) + \tau_2(6) + \tau_2(9) + \tau_2(12) + \tau_2(18) + \tau_2(36) =$

   $1 + 3 + 3 + 6 + 9 + 6 + 18 + 18 + 36 = 100$.

4. $\tau_3(36) = \tau_3(4)\tau_3(9) = \tau_3(2^2)\tau_3(3^2) = \binom{2+3}{3}\binom{2+3}{3} = \binom{5}{3}\binom{5}{3} = 10 \cdot 10 = 100$.

5. $\sum_{k=0}^{3}(-1)^k \binom{4}{k}\tau_{3-k}(6) = \binom{4}{0}\tau_3(6) - \binom{4}{1}\tau_2(6) + \binom{4}{2}\tau_1(6) - \binom{4}{3}\tau_0(6) = 16 - 36 + 24 - 4 = 0$.

Chapter 11.

1. Corollary 11.9 states that $B_m(\lambda) = \sum_{k=1}^{m}\binom{m}{k}\lambda_k \int_0^1 B_{m-k}(\lambda, u)\,du$ where $B_0 = 1$. So, for m = 5, we have

$B_5(\lambda) = \sum_{k=1}^{5}\binom{5}{k}\lambda_k \int_0^1 B_{5-k}(\lambda, u)\,du = \binom{5}{1}\lambda_1\int_0^1 B_4(\lambda, u)\,du + \binom{5}{2}\lambda_2\int_0^1 B_3(\lambda, u)\,du + \binom{5}{3}\lambda_3\int_0^1 B_2(\lambda, u)\,du +$

$\binom{5}{4}\lambda_4\int_0^1 B_1(\lambda, u)\,du + \binom{5}{5}\lambda_5\int_0^1 B_0(\lambda, u)\,du$. Now follow Example 11.10 to calculate $B_5(\lambda)$.

2. The 3 in $B_3(\lambda)$ means that we have 3 distinct objects. The coefficient of λ_3, in $B_3(\lambda) = \lambda_3 + 3\lambda_1\lambda_2 + \lambda_1^3$, is 1, and that means that there is one distribution of the objects into one box that contains all three objects. The term $3\lambda_1\lambda_2$ means there are 3 distributions of the objects into two boxes such that one box contains one object (indicated by λ_1) and another box contains two objects (indicated by λ_2). And there is one distribution such that one box contains one object, another box contains one object, and another box contains one object (indicated by the term λ_1^3.

3. If the specification number n represents k distinct objects, then the object distribution polynomial $g_n(\lambda, t) = B_k(\lambda, t)$.

4. If we let $\lambda_j = 1$ for all j in $B_m(\lambda_1, \lambda_2, \ldots, \lambda_m)$, we get $B(m)$, the mth Bell number. For example, $B_3(\lambda) = B_3(\lambda_1, \lambda_2, \lambda_3) = \lambda_3 + 3\lambda_1\lambda_2 + \lambda_1^3$. So, $B_3(1, 1, 1) = 5$.

5. We will simply outline the solution. We want to calculate $g_{18}(\lambda, t)$. Here, the specification number for the set of objects is $n = 18 = 2 \cdot 3^2$. The divisors d of 18 are $\{1, 2, 3, 6, 9, 18\}$, but we only want to use the divisors, 1, 2, 3, 6, and 9. By Theorem 11.11, we have the recurrence relation

$g_{18}(\lambda, t) = \sum_{\substack{d\mid n \\ d\neq n}}\int_0^1 \frac{1}{x}g_d(\lambda, x)P_{\frac{18}{d}}(\lambda, x)\,dx$. Now follow Example 11.12.

Chapter 12.

1. For two red balls and two blue balls, we can let the specification number be $n = 2^2 \cdot 3^2 = 36$. We want to calculate $f_{36}(t)$. We will simply set up the problem. By Theorem 12.5, we have

$$f_{36}(t) = \sum_{\substack{d|36 \\ d \neq 36}} \int_0^t \frac{1}{x} f_d(x) P_{\frac{36}{d}}(p, x) dx.$$ Follow the calculations in Example 12.6.

2. By Theorem 12.8, we have $\varphi(r, t) = r \cdot \varphi(r-1, t) + r \cdot \int_0^t \varphi(r-1, x) dx$ with $\varphi(1, t) = t$. Then

$\varphi(2, t) = 2 \cdot \varphi(1, t) + 2 \cdot \int_0^t \varphi(1, x) dx = 2t + 2 \cdot \int_0^t x \, dx = 2t + t^2$. Now that we know $\varphi(2, t)$, we can

calculate $\varphi(3, t)$: $\varphi(3, t) = 3 \cdot \varphi(2, t) + 3 \cdot \int_0^t \varphi(2, x) dx = 3 \cdot (2t + t^2) + 3 \cdot \int_0^t (2x + x^2) dx = 6t + 6t^2 + t^3$.

Likewise, we have $\varphi(4, t) = 4 \cdot \varphi(3, t) + 4 \cdot \int_0^t \varphi(3, x) dx = 24t + 36t^2 + 12t^3 + t^4$.

3. Corollary 12.10 states that $S_O(6, 4) = 6 \cdot S_O(5, 4) + \frac{6}{4} \cdot S_O(5, 3)$. Now we can use Corollary 12.10 again to compute $S_O(5, 4)$ and $S_O(5, 3)$. Continue using Corollary 12.10 and back-substitution, then simplify the result to obtain $S_O(6, 4) = 300$.

4. The partitions of 7 into 3 parts are $[1^2 5^1]$, $[1^1 2^1 4^1]$, $[1^1 3^2]$, and $[2^2 3^1]$. So, by Corollary 12.13, we

have $\binom{3}{2, 1} + \binom{3}{1, 1, 1} + \binom{3}{1, 2} + \binom{3}{2, 1} = \frac{3!}{1!2!} + \frac{3!}{1!1!1!} + \frac{3!}{1!2!} + \frac{3!}{1!2!} = 3 + 6 + 3 + 3 = 15$. Also,

$\binom{r-1}{k-1} = \binom{7-1}{3-1} = \binom{6}{2} = \frac{6!}{2!4!} = 15$.

Chapter 13.

1. For $n = 6$, we have $g_{36}(t) * g_{36}(-t) = \sum_{d|36} g_d(t) g_{\frac{36}{d}}(-t) = g_1(t) g_{36}(-t) + g_2(t) g_{18}(-t) + g_3(t) g_{12}(-t) +$

$g_4(t) g_9(-t) + g_6(t) g_6(-t) + g_9(t) g_4(-t) + g_{12}(t) g_3(-t) + g_{18}(t) g_2(-t) + g_{36}(t) g_1(-t) =$

$1 \cdot (-t + 4t^2 - 3t^3 + t^4) + t \cdot (-t + 2t^2 - t^3) + t \cdot (-t + 2t^2 - t^3) + (t + t^2)(-t + t^2) + (t + t^2)(-t + t^2) +$

$(t + t^2)(-t + t^2) + (t + 2t^2 + t^3)(-t) + (t + 2t^2 + t^3)(-t) + (t + 4t^2 + 3t^3 + t^4) \cdot 1 = t^2 + t^4$. But $t^2 + t^4 =$

$g_6(t^2)$ since $g_6(t) = t + t^2$.

2. Solve this problem exactly as we did above, in problem 1, but look up the delta-polynomials in Table 5.1.

3. We want to show that $\varepsilon_2(t^2) = 0$. By Theorem 13.4, we have $g_2(t) * g_2(-t) = \varepsilon_2(t^2)$. Thus, we have $g_2(t) * g_2(-t) = \sum_{d|2} g_d(t) g_{\frac{2}{d}}(-t) = g_1(t)g_2(-t) + g_2(t)g_1(-t) = 1 \cdot (-t) + t \cdot 1 = 0$. If the specification number for the object set is $n = 2$, which we can think of as a single red ball because 2 is a single prime number, then $\varepsilon_2(u)$ is the object distribution polynomial that counts the number of ways to distribute one red ball into identical boxes, with no box empty, such that each box has even color occupancy. However, this is impossible, because we only have one red ball, not an even number of red balls. Therefore, we expect $\varepsilon_2(u) = \varepsilon_2(t^2) = 0$.

4. For $n = 4$, we have $g_4(t) * g_4(-t) = \sum_{d|4} g_d(t) g_{\frac{4}{d}}(-t) = g_1(t)g_4(-t) + g_2(t)g_2(-t) + g_4(t)g_1(-t) =$

$1 \cdot (-t + t^2) + t \cdot (-t) + (t + t^2) \cdot 1 = t^2 = g_2(t^2)$ as expected by Theorem 13.6. For $n = 12$, we have

$g_{12}(t) * g_{12}(-t) = \sum_{d|12} g_d(t) g_{12}(-t) = g_1(t)g_{12}(-t) + g_2(t)g_6(-t) + g_3(t)g_4(-t) + g_4(t)g_3(-t) +$

$g_6(t)g_2(-t) + g_{12}(t)g_1(-t) = 1 \cdot (-t + 2t^2 - t^3) + t \cdot (-t + t^2) + t \cdot (-t + t^2) + (t + t^2)(-t) + (t + t^2)(-t)$

$+ (t + 2t^2 + t^3) \cdot 1 = 0$, as expected by Theorem 13.6.

Chapter 14.

1. Label the vertices of a regular pentagon as 1, 2, 3, 4, and 5 clockwise around the pentagon. The group G of symmetries of the regular pentagon consists of the rigid mappings of the pentagon into itself. These mappings can be represented as a set of permutations of the labeled vertices. There are five rotations and five flips. (The rotation through zero degrees is just the identity mapping, ι.) We can represent the rotations through theta degrees as ρ_θ, and we can represent the flips about an axis through vertex x as f_x. Doing so, we have the following rigid mappings of the regular pentagon:

$$\iota = \begin{pmatrix} 1 & 2 & 3 & 4 & 5 \\ 1 & 2 & 3 & 4 & 5 \end{pmatrix} = (1)(2)(3)(4)(5) = x_1^5$$

$$\rho_{72°} = \begin{pmatrix} 1 & 2 & 3 & 4 & 5 \\ 2 & 3 & 4 & 5 & 1 \end{pmatrix} = (12345) = x_5$$

$$\rho_{144°} = \begin{pmatrix} 1 & 2 & 3 & 4 & 5 \\ 3 & 4 & 5 & 1 & 2 \end{pmatrix} = (13524) = x_5$$

$$\rho_{216°} = \begin{pmatrix} 1 & 2 & 3 & 4 & 5 \\ 4 & 5 & 1 & 2 & 3 \end{pmatrix} = (14253) = x_5$$

$$\rho_{288°} = \begin{pmatrix} 1 & 2 & 3 & 4 & 5 \\ 5 & 1 & 2 & 3 & 4 \end{pmatrix} = (15432) = x_5$$

$$f_1 = \begin{pmatrix} 1 & 2 & 3 & 4 & 5 \\ 1 & 5 & 4 & 3 & 2 \end{pmatrix} = (1)(25)(34) = x_1 x_2^2$$

$$f_2 = \begin{pmatrix} 1 & 2 & 3 & 4 & 5 \\ 3 & 2 & 1 & 5 & 4 \end{pmatrix} = (13)(2)(45) = x_1 x_2^2$$

$$f_3 = \begin{pmatrix} 1 & 2 & 3 & 4 & 5 \\ 5 & 4 & 3 & 2 & 1 \end{pmatrix} = (15)(24)(3) = x_1 x_2^2$$

$$f_4 = \begin{pmatrix} 1 & 2 & 3 & 4 & 5 \\ 2 & 1 & 5 & 4 & 3 \end{pmatrix} = (12)(35)(4) = x_1 x_2^2$$

$$f_5 = \begin{pmatrix} 1 & 2 & 3 & 4 & 5 \\ 4 & 3 & 2 & 1 & 5 \end{pmatrix} = (14)(23)(5) = x_1 x_2^2$$

Since the group G of rotations and flips contains 10 elements, or permutations, the order of G is 10: $|G| = 10$. Therefore, the cycle index for the rigid mappings of a regular pentagon into itself is

$$P_G = \frac{1}{10}\left(x_1^5 + 4x_5 + 5x_1 x_2^2\right).$$

2. If we exclude the flips, the subgroup H of G consists only of rotations and has cycle idex:

$$P_G = \frac{1}{5}\left(x_1^5 + 4x_5\right).$$

3. In Example 14.5, we have the rotations $\pi_0 = x_1^3$, $\pi_{120} = x_3$, and $\pi_{240} = x_3$. The flips are

$$f_1 = \begin{pmatrix} 1 & 2 & 3 \\ 1 & 3 & 2 \end{pmatrix} = (1)(23) = x_1 x_2, \quad f_2 = \begin{pmatrix} 1 & 2 & 3 \\ 3 & 2 & 1 \end{pmatrix} = (2)(13) = x_1 x_2, \text{ and}$$

$$f_3 = \begin{pmatrix} 1 & 2 & 3 \\ 2 & 1 & 3 \end{pmatrix} = (3)(12) = x_1 x_2.$$ So, the cycle index for both rotations and flips of an equilateral triangle is $P_G = \frac{1}{6}\left(x_1^3 + 2x_3 + 3x_1 x_2\right).$

Bibliography

[1] J. Riordan, *Introduction to Combinatorial Analysis*, Dover edition, Dover Publications, Inc., Mineola, New York, 2002.

[2] C. Charlambides, *Enumerative Combinatorics*, Chapman & Hall/CRC, Boca Raton, Florida, 2002.

[3] P. A. MacMahon, *Combinatory Analysis*, Vol. 1, Cambridge University Press, London, 1915.

[4] G. E. Andrews, *The Theory of Partitions*, Cambridge University Press, New York, 1984.

[5] N. L. Biggs, *Discrete Mathematics*, Second edition, Oxford University Press, New York, 2002.

[6] N. J. A. Sloane and S. Plouffe, *The Encyclopedia of Integer Sequences*, Academic Press, San Diego, 1995.

[7] G. H. Hardy and E. M. Wright, *An Introduction to the Theory of Numbers*, Fifth edition, Oxford University Press, New York, 1979.

[8] W. Dunham, *Euler, The Master of Us All*, The Mathematical Association of America, Washington, D. C., 1999.

Index

Algebraic identities, 223
All parts divisible by m, 136
Arbitrary specification, 47
Average number of boxes per distribution, 51
Bell numbers, 45, 113, 115
Bibliography, 271
Boxes of several types, 149
Boxes of two colors, 232
Building even occupancy polynomials, 121
Building general occupancy polynomials, 90
Cauchy-Schwartz inequality, 33
Cauchy-type product rule, 20
Combinatorial number theory, ix
Convergence properties, 29, 30
Convolution product, 20
Counting function, 29
Cycle index, 238
Dirichlet series generating functions, 78
Distinct boxes, 161
Distinct objects in distinct boxes, 188
Distinct occupancies, 65
Distinct occupancy generating function, 84
Distribution and occupancy, x
Enumeration formulas, 11
Enumeration problems, 1
Equivalent problems, 24

Euler's generating function, 17
Euler's theorem, 83
Even color occupancies, 93
Exponential Bell partition polynomials, 191
Exponential formulas, 81
Exponential relations for partition functions, 80
Factorizations of a positive integer, 25
Factorizations, 23
Ferris wheel problem, 240
Functional integral equation, 181
Fundamental Theorem of Arithmetic, 16
General convolution product identity, 233
General theorem on integer partitions, 133
Generalization of Euler's partition theorem, 87, 89
Generalization of Polya's enumeration theorem, 237, 241
Generalized exponential Bell partition polynomial, 195
Generalized partitions, 113
Generating function for distinct boxes, 175, 184
Generating function for even color occupancy, 95
Group theory, 237
Growth rate, 32

Holy Grail, 11
Identical objects, 73, 102
Infinite series, 29
Instantaneous slope, 49
Integer partition formula, 12
Integer Power Polynomial, 52
Integral recurrence formula, 50, 73
Iversonian convention, 106
Linear independence, 21
Logarithmic generating function, 19
Logarithms, 18
Mobius function, 140
Mobius inversion formula, 77
Multiplicative inverse, 84
No parts divisible by m, 136
Number theoretic combinatorics, ix
Number theory, ix
Number-theoretic method, 13
Object distribution polynomial, 14
Object distribution polynomials, 251
Objects in identical boxes, 24, 44
Ordered occupancies, 205
Ordered Stirling numbers of second kind, 216
Partitions into distinct parts, 74
Partitions into even parts, 84
Permutation function, 207
Permutation integer power polynomial, 208
Polya's enumeration theorem, 238
Polynomial identities, 125
Properties of generalized divisor function, 168
Recurrence relations, 39, 68, 97

Relationships among generating functions, 127
Research problems, 257
Riemann zeta function, 35
Second integral recurrence formula, 53
Simple formula, 11
Solutions to exercises, 259
Specification number, 17
Square Root Theorem, 119
Stirling numbers of second kind, 7, 8, 45
Sum of divisors, 79
Sum of even divisors, 80
Sum of odd divisors function, 75
Sum of odd divisors, 79
Symmetric object sets, 139
Triple color multiplicity, 125, 126
Upper bound, 35
Upper summation identity, 187
Vector partitions, 26

www.ingramcontent.com/pod-product-compliance
Lightning Source LLC
Chambersburg PA
CBHW080905170526
45158CB00008B/1999